WILEY-INTERSCIENCE
SERIES IN DISCRETE MATHEMATICS AND OPTIMIZATION

AARTS AND KORST
Simulated Annealing and Boltzmann Machines: A Stochastic Approach to Combinatorial Optimization and Neural Computing

AHLSWEDE AND WEGENER
Search Problems

ALON, SPENCER, AND ERDŐS
The Probabilistic Method

ANDERSON AND NASH
Linear Programming in Infinite-Dimensional Spaces: Theory and Application

BAZARAA, JARVIS AND SHERALI
Linear Programming and Network Flows

BEALE
Introduction to Optimization

COFFMAN AND LUEKER
Probabilistic Analysis of Packing and Partitioning Algorithms

DINITZ AND STINSON
Contemporary Design Theory: A Collection of Surveys

GONDRAN AND MINOUX
Graphs and Algorithms
(*Translated by S. Vajda*)

GOULDEN AND JACKSON
Combinatorial Enumeration

GRAHAM, ROTHSCHILD, AND SPENCER
Ramsey Theory
Second Edition

GROSS AND TUCKER
Topological Graph Theory

HALL
Combinatorial Theory
Second Edition

LAWLER, LENSTRA, RINNOOY KAN, AND SHMOYS, EDITORS
The Traveling Salesman Problem: A Guided Tour of Combinatorial Optimization

MAHMOUD
Evolution of Random Search Trees

MARTELLO AND TOTH
Knapsack Problems: Algorithms and Computer Implementations

MINC
Nonnegative Matrices

MINOUX
Mathematical Programming: Theory and Algorithms
(*Translated by S. Vajda*)

MIRCHANDANI AND FRANCIS, EDITORS
Discrete Location Theory

NEMHAUSER AND WOLSEY
Integer and Combinatorial Optimization

NEMIROVSKY AND YUDIN
Problem Complexity and Method Efficiency in Optimization
(*Translated by E. R. Dawson*)

PLESS
Introduction to the Theory of Error-Correcting Codes
Second Edition

SCHRIJVER
Theory of Linear and Integer Programming

TOMESCU
Problems in Combinatorics and Graph Theory
(*Translated by R. A. Melter*)

TUCKER
Applied Combinatorics
Second Edition

The Probabilistic Method

The Probabilistic Method

NOGA ALON
Department of Mathematics
Raymond and Beverly Sackler Faculty of Exact Sciences
Tel Aviv University
Tel Aviv, Israel

JOEL H. SPENCER
Courant Institute of Mathematical Sciences
New York University
New York, New York

With an appendix on open problems by

PAUL ERDŐS
Mathematical Institute of the Hungarian Academy of Sciences
Budapest, Hungary

A Wiley-Interscience Publication
JOHN WILEY & SONS, INC.
New York • Chichester • Brisbane • Toronto • Singapore

A NOTE TO THE READER
This book has been electronically reproduced from digital information stored at John Wiley & Sons, Inc. We are pleased that the use of this new technology will enable us to keep works of enduring scholarly value in print as long as there is a reasonable demand for them. The content of this book is identical to previous printings.

In recognition of the importance of preserving what has been written, it is a policy of John Wiley & Sons, Inc., to have books of enduring value published in the United States printed on acid-free paper, and we exert our best efforts to that end.

Library of Congress Cataloging in Publication Data:

Alon, Noga.
The probabilistic method/Noga Alon, Joel H. Spencer: with an appendix on open problems by Paul Erdős.
p. cm.—(Wiley Interscience series in discrete mathematics and optimization)
Includes bibliographical references and index.
ISBN 0-471-53588-5
1. Combinatorial analysis. 2. Probabilities. I. Spencer, Joel H. II. Erdős, Paul. III. Title. IV. Series.
QA164.A46 1991
511′.6–dc20 91-13119
CIP

Printed and bound in the United States of America

10 9 8 7 6 5 4

To Nurit and Mary Ann

Preface

The probabilistic method has recently been developed intensively and became one of the most powerful and widely used tools applied in combinatorics. One of the major reasons for this rapid development is the important role of randomness in theoretical computer science, a field that is recently the source of many intriguing combinatorial problems.

The interplay between discrete mathematics and computer science suggests an algorithmic point of view in the study of the probabilistic method in combinatorics, and this is the approach we tried to adopt in this book. The text thus includes a discussion of algorithmic techniques together with a study of the classical method as well as the modern tools applied in it. The first part of the book contains a description of the tools applied in probabilistic arguments, including the basic techniques that use expectation and variance, as well as the more recent applications of martingales and correlation inequalities. The second part includes a study of various topics in which probabilistic techniques have been successful. This part contains chapters on discrepancy and random graphs, as well as on several areas in theoretical computer science: circuit complexity, computational geometry, and derandomization of randomized algorithms. Scattered between the chapters are gems described under the heading "The Probabilistic Lens." These are elegant proofs that are not necessarily related to the chapters after which they appear and can be usually read separately.

The basic probabilistic method can be described as follows: in order to prove the existence of a combinatorial structure with certain properties, we construct an appropriate probability space and show that a randomly chosen element in this space has the desired properties with positive probability. This method has been initiated by Paul Erdős, who contributed so much to its development over the last 45 years that it seems appropriate to call it "the Erdős method." His contribution can be measured not only by his numerous deep results in the subject, but also by his many intriguing problems and conjectures that stimulated a big portion of the research in the area. It is therefore very pleasing that we can close this book with an "open problems" appendix by Paul Erdős, the founder of the probabilistic method.

It seems impossible to write an encyclopedic book on the probabilistic method; too many recent interesting results apply probabilistic arguments, and we do not even try to mention all of them. Our emphasis is on methodology, and we thus try to describe the ideas, and not always to give the best possible results if these are too technical to allow a clear presentation. Many of the results are asymptotic, and we use the standard asymptotic notation: for two functions f and g, we write $f = O(g)$ if $f \leq c_1 g + c_2$ for all possible values of the variables of the two functions, where c_1, c_2 are absolute constants. We write $f = \Omega(g)$ if $g = O(f)$ and $f = \Theta(g)$ if $f = O(g)$ and $f = \Omega(g)$. If the limit of the ratio f/g tends to 0 as the variables of the functions tend to infinity, we write $f = o(g)$. Finally, $f \sim g$ denotes that $f = (1 + o(1))g$, i.e., that f/g tends to 1 when the variables tend to infinity.

It is a special pleasure to thank our wives, Nurit and Mary Ann. Their patience, understanding, and encouragement have been key ingredients in the success of this enterprise.

NOGA ALON
JOEL H. SPENCER

Contents

The Probabilistic Method

Part I

METHODS

1

The Basic Method

1. THE PROBABILISTIC METHOD

The probabilistic method is a powerful tool in tackling many problems in discrete mathematics. Roughly speaking, the method works as follows: Trying to prove that a structure with certain desired properties exists, one defines an appropriate probability space of structures and then shows that the desired properties hold in this space with positive probability. The method is best illustrated by examples. Here is a simple one. The *Ramsey number* $R(k,\ell)$ is the smallest integer n such that in any 2-coloring of the edges of a complete graph on n vertices K_n by red and blue, there either is a red K_k (i.e., a complete subgraph on k vertices, all of whose edges are colored red), or there is a blue K_ℓ. Ramsey (1930) showed that $R(k,\ell)$ is finite for any two integers k and ℓ. Let us obtain a lower bound for the diagonal Ramsey numbers $R(k,k)$.

Proposition 1.1. *If* $\binom{n}{k} \cdot 2^{1-\binom{k}{2}} < 1$, *then* $R(k,k) > n$. *Thus* $R(k,k) > 2^{k/2}$ *for all* $k \geq 3$.

Proof. Consider a random 2-coloring of the edges of K_n obtained by coloring each edge independently either red or blue, where each color is equally likely. For any fixed set R of k vertices, let A_R be the event that the induced subgraph of K_n on R is *monochromatic* (i.e., that either all its edges are red or they are all blue). Clearly, $\Pr(A_R) = 2^{1-\binom{k}{2}}$. Since there are $\binom{n}{k}$ possible choices for R, the probability that at least one of the events A_R occurs is at most $\binom{n}{k}2^{1-\binom{k}{2}} < 1$. Thus, with positive probability, no event A_R occurs and there is a 2-coloring of K_n without a monochromatic K_k, that is, $R(k,k) \geq n$. Note that if $k \geq 3$ and we take $n = \lfloor 2^{k/2} \rfloor$, then

$$\binom{n}{k} 2^{1-\binom{k}{2}} < \frac{2^{1+k/2}}{k!} \cdot \frac{n^k}{2^{k^2/2}} < 1,$$

and hence $R(k,k) > 2^{k/2}$ for all $k \geq 3$. ■

This simple example demonstrates the essence of the probabilistic method. To prove the existence of a good coloring we do not present one explicitly, but rather show, in a nonconstructive way, that it exists. This example appeared in a paper of P. Erdős from 1947. Although Szele applied the probabilistic method to another combinatorial problem, mentioned in Chapter 2, already in 1943, Erdős was certainly the first one who understood the full power of this method and has been applying it successfully over the years to numerous problems. One can, of course, claim that the probability is not essential in the proof given above. An equally simple proof can be described by counting; we just check that the total number of 2-colorings of K_n is bigger than the number of those containing a monochromatic K_k. Moreover, since the vast majority of the probability spaces considered in the study of combinatorial problems are finite spaces, this claim applies to most of the applications of the probabilistic method in discrete mathematics. Theoretically, this is, indeed, the case. However, in practice, the probability is essential. It would be hopeless to replace the applications of many of the tools appearing in this book, including, for example, the second moment method, the Lovász local lemma, and the concentration via martingales by counting arguments, even when these are applied to finite probability spaces.

The probabilistic method has an interesting algorithmic aspect. Consider, for example, the proof of Proposition 1.1 that shows that there is an edge 2-coloring of K_n without a monochromatic $K_{2\log_2 n}$. Can we actually find such a coloring? This question, as asked, may sound ridiculous; the total number of possible colorings is finite, so we can try them all until we find the desired one. However, such a procedure may require $2^{\binom{n}{2}}$ steps; an amount of time that is exponential in the size ($= \binom{n}{2}$) of the problem. Algorithms whose running time is more than polynomial in the size of the problem are usually considered unpractical. The class of problems that can be solved in polynomial time, usually denoted by P (see, e.g., Aho, Hopcroft, and Ullman [1974]), is, in a way, the class of all solvable problems. In this sense, the exhaustive search approach suggested above for finding a good coloring of K_n is not acceptable, and this is the reason for our remark that the proof of Proposition 1.1 is nonconstructive; it does not supply a constructive, efficient, and deterministic way of producing a coloring with the desired properties. However, a closer look at the proof shows that, in fact, it can be used to produce, effectively, a coloring that is very likely to be good. This is because for large k, if $n = \lfloor 2^{k/2} \rfloor$ then

$$\binom{n}{k} \cdot 2^{1-\binom{k}{2}} < \frac{2^{1+k/2}}{k!}\left(\frac{n}{2^{k/2}}\right)^k \leq \frac{2^{1+k/2}}{k!} \ll 1.$$

Hence a random coloring of K_n will most probably not contain a monochromatic $K_{2\log n}$. This means that if, for some reason, we *must* present a 2-coloring of the edges of K_{1024} without a monochromatic K_{20}, we can simply produce a random 2-coloring by flipping a fair coin $\binom{1024}{2}$ times. We can then hand the resulting coloring safely; the probability that it contains a monochromatic K_{20} is less than $2^{11}/20!$, probably much smaller than our chances of making a mistake in any rigorous proof that a certain coloring is good! Therefore, in some cases, the probabilistic, nonconstructive method does supply effective probabilistic algorithms. Moreover, these algorithms can sometimes be converted into deterministic ones. This topic is discussed in some detail in Chapter 15.

The probabilistic method is a powerful tool in combinatorics and in graph theory. It is also extremely useful in number theory and in combinatorial geometry. More recently, it has been applied in the development of efficient algorithmic techniques and in the study of various computational problems. In the rest of this chapter we present several simple examples that demonstrate some of the broad spectrum of topics in which this method is helpful. More complicated examples, involving various more delicate probabilistic arguments, appear in the rest of the book.

2. GRAPH THEORY

A *tournament* on a set V of n players is an orientation $T = (V, E)$ of the edges of the complete graph on the set of vertices V. Thus for every two distinct elements x and y of V either (x,y) or (y,x) is in E, but not both. The name tournament is natural, since one can think of the set V as a set of players in which each pair participates in a single match, where (x,y) is in the tournament iff x beats y. We say that T has the property S_k if for every set of k players there is one who beats them all. For example, a directed triangle $T_3 = (V, E)$, where $V = \{1,2,3\}$ and $E = \{(1,2),(2,3),(3,1)\}$, has S_1. Is it true that for every finite k there is a tournament T (on more than k vertices) with the property S_k? As shown by Erdős (1963a), this problem, raised by Schütte, can be solved almost trivially by applying probabilistic arguments. Moreover, these arguments even supply a rather sharp estimate for the minimum possible number of vertices in such a tournament. The basic (and natural) idea is that if n is sufficiently large, as a function of k, then a *random* tournament on the set $V = \{1,\ldots,n\}$ of n players is very likely to have property S_k. By a random tournament we mean here a tournament T on V obtained by choosing, for each $1 \le i < j \le n$, independently, either the edge (i,j) or the edge (j,i), where each of these two choices is equally likely. Observe that in this manner, all the $2^{\binom{n}{2}}$ possible tournaments on V are equally likely, that is, the probability space considered is symmetric. It is worth noting that we often use in applications symmetric probability spaces. In these cases, we shall sometimes refer to an element of the space as a *random element*, without describing explicitly the probability distribution. Thus, for example, in the proof of Proposition 1.1 random 2-edge-colorings of K_n were considered, i.e., all possible colorings were equally likely. Similarly, in the proof of the next simple result, we study random tournaments on V.

Theorem 2.1. *If $\binom{n}{k}(1-2^{-k})^{n-k} < 1$, then there is a tournament on n vertices that has the property S_k.*

Proof. Consider a random tournament on the set $V = \{1,\ldots,n\}$. For every fixed subset K of size k of V, let A_K be the event that there is no vertex that beats all the members of K. Clearly, $\Pr(A_K) = (1-2^{-k})^{n-k}$. This is because for each fixed vertex $v \in V - K$, the probability that v does not beat all the members of K is $1-2^{-k}$, and all these $n-k$ events corresponding to the various possible choices of v are independent. It follows that

$$\Pr\left(\bigcup_{\substack{K\subset V\\|K|=k}} A_K\right) \le \sum_{\substack{K\subset V\\|K|=k}} \Pr(A_K) = \binom{n}{k}(1-2^{-k})^{n-k} < 1.$$

Therefore, with positive probability no event A_K occurs, i.e., there is a tournament on n vertices that has the property S_k. ■

Let $f(k)$ denote the minimum possible number of vertices of a tournament that has the property S_k. Since $\binom{n}{k} < (en/k)^k$ and $(1-2^{-k})^{n-k} < e^{-(n-k)/2^k}$, Theorem 2.1 implies that $f(k) \le k^2 \cdot 2^k \cdot (\ln 2)(1+o(1))$. It is not too difficult to check that $f(1) = 3$ and $f(2) = 7$. As proved by Szekeres (cf. Moon [1968]), $f(k) \ge c_1 \cdot k \cdot 2^k$. Can one find an explicit construction of tournaments with at most c_2^k vertices having property S_k? Such a construction is known, but is not trivial; it is described in Chapter 9.

A *dominating set* of an undirected graph $G = (V,E)$ is a set $U \subseteq V$ such that every vertex $v \in V - U$ has at least one neighbor in U.

Theorem 2.2. *Let $G = (V,E)$ be a graph on n vertices, with minimum degree $\delta > 1$. Then G has a dominating set of at most $n[1+\ln(\delta+1)]/(\delta+1)$ vertices.*

Proof. Put $p = \ln(\delta+1)/(\delta+1)$ and let us pick, randomly and independently, each vertex of V with probability p. Let X be the (random) set of all vertices picked and let $Y = Y_X$ be the random set of all vertices in $V - X$ that do not have any neighbor in X. The expected value of $|X|$ is clearly np. For each fixed vertex $v \in V$, $\Pr(v \in Y) = \Pr(v$ and its neighbors are not in $X) \le (1-p)^{\delta+1}$. Since the expected value of a sum of random variables is the sum of their expectations (even if they are not independent) and since the random variable $|Y|$ can be written as a sum of n indicator random variables χ_v $(v \in V)$, where $\chi_v = 1$ if $v \in Y$ and $\chi_v = 0$ otherwise, we conclude that the expected value of $|X| + |Y|$ is at most

$$np + n(1-p)^{\delta+1} = n\frac{\ln(\delta+1)}{\delta+1} + n\left(1 - \frac{\ln(\delta+1)}{\delta+1}\right)^{\delta+1} \le n\frac{1+\ln(\delta+1)}{\delta+1}.$$

Consequently, there is at least one choice of $X \subseteq V$ such that

$$|X| + |Y_X| \le n\frac{1+\ln(\delta+1)}{\delta+1}.$$

The set $U = X \cup Y_X$ is clearly a dominating set of G whose cardinality is at most $n[1+\ln(\delta+1)]/(\delta+1)$. ■

Two simple but important ideas are incorporated in the last proof. The first is the linearity of expectation; many applications of this simple, yet powerful principle appear in Chapter 2. The second is, maybe, more subtle, and is an example of the "alteration" principle which is discussed in Chapter 3. The random choice did not supply the required dominating set U immediately; it only supplied the set X, which has to be altered a little (by adding to it the set Y_X) to provide the required dominating set.

It can be easily deduced from the results in Alon (1990a) that the bound in Theorem 2.2 is nearly optimal. A nonprobabilistic, algorithmic, proof of this theorem can be obtained by choosing the vertices for the dominating set one by one, when in each step a vertex that covers the maximum number of yet uncovered vertices is picked. Indeed, for each vertex v, denote by $C(v)$ the set consisting of v together with all its neighbors. Suppose that during the process of picking vertices the number of vertices u that do not lie in the union of the sets $C(v)$ of the vertices chosen so far is r. By the assumption, the sum of the cardinalities of the sets $C(u)$ over all such uncovered vertices u is at least $r(\delta+1)$, and hence, by averaging, there is a vertex v that belongs to at least $r(\delta+1)/n$ such sets $C(u)$. Adding this v to the set of chosen vertices we observe that the number of uncovered vertices is now at most $r(1-(\delta+1)/n)$. It follows that in each iteration of the above procedure the number of uncovered vertices decreases by a factor of $1-(\delta+1)/n$ and hence after $[n\ln(\delta+1)]/(\delta+1)$ steps there will be at most $n/(\delta+1)$ yet uncovered vertices which can now be added to the set of chosen vertices to form a dominating set of size at most the one in the conclusion of Theorem 2.2.

Combining this with some of the ideas of Matula (1987), we can obtain a very efficient algorithm to decide if a given undirected graph on n vertices is, say, $(n/2)$-edge connected. A *cut* in a graph $G=(V,E)$ is a partition of the set of vertices V into two nonempty disjoint sets $V=V_1\cup V_2$. If $v_1\in V_1$ and $v_2\in V_2$, we say that the cut *separates* v_1 and v_2. The *size* of the cut is the number of edges of G having one end in V_1 and another end in V_2. In fact, we sometimes identify the cut with the set of these edges. The *edge connectivity* of G is the minimum size of a cut of G. The following lemma is due to Matula.

Lemma 2.3. *Let $G=(V,E)$ be a graph with minimum degree δ and let $V=V_1\cup V_2$ be a cut of size smaller than δ in G. Then every dominating set U of G has vertices in V_1 and in V_2.*

Proof. Suppose this is false and $U\subseteq V_1$. Choose, arbitrarily, a vertex $v\in V_2$ and let $v_1,v_2,\ldots,v_\delta$ be δ of its neighbors. For each i, $1\le i\le\delta$, define an edge e_i of the given cut as follows: if $v_i\in V_1$ then $e_i=\{v,v_i\}$, otherwise, $v_i\in V_2$ and since U is dominating there is at least one vertex $u\in U$ such that $\{u,v_i\}$ is an edge; take such a u and put $e_i=\{u,v_i\}$. The δ edges $e_1,\ldots,e_\delta$ are all distinct and all lie in the given cut, contradicting the assumption that its size is less than δ. This completes the proof. ■

Let $G = (V, E)$ be a graph on n vertices, and suppose we wish to decide if G is $n/2$ edge-connected, that is, if its edge connectivity is at least $n/2$. Matula showed, by applying Lemma 2.3, that this can be done in time $O(n^3)$. By the remark following the proof of Theorem 2.2, we can slightly improve it and get an $O(n^{8/3}\log n)$ algorithm as follows. We first check if the minimum degree δ of G is at least $n/2$. If not, G is not $n/2$ edge-connected, and the algorithm ends. Otherwise, by Theorem 2.2, there is a dominating set $U = \{u_1, \ldots, u_k\}$ of G, where $k = O(\log n)$, and it can in fact be found in $O(n^2)$ time. We now find, for each i, $2 \le i \le k$, the minimum size s_i of a cut that separates u_1 from u_i. Each of these problems can be solved by solving a standard network flow problem in time $O(n^{8/3})$, (see, e.g., Tarjan [1983].) By Lemma 2.3, the edge connectivity of G is simply the minimum between δ and $\min_{2 \le i \le k} s_i$. The total time of the algorithm is $O(n^{8/3}\log n)$, as claimed.

3. COMBINATORICS

A *hypergraph* is a pair $H = (V, E)$, where V is a finite set whose elements are called *vertices* and E is a family of subsets of V, called *edges*. It is *n-uniform* if each of its edges contains precisely n vertices. We say that H has *property* B, or that it is *2-colorable* if there is a 2-coloring of v such that no edge is monochromatic. Let $m(n)$ denote the minimum possible number of edges of an n-uniform hypergraph that does not have property B.

Proposition 3.1 (Erdős [1963b]). *Every n-uniform hypergraph with less than 2^{n-1} edges has property B. Therefore $m(n) \ge 2^{n-1}$.*

Proof. Let $H = (V, E)$ be an n-uniform hypergraph with less than 2^{n-1} edges. Color V randomly by two colors. For each edge $e \in E$, let A_e be the event that e is monochromatic. Clearly, $\Pr(A_e) = 2^{1-n}$. Therefore

$$\Pr\left(\bigcup_{e \in E} A_e\right) \le \sum_{e \in E} \Pr(A_e) < 1,$$

and there is a 2-coloring without monochromatic edges. ■

In Chapter 3, Section 5, we present a more delicate argument, due to J. Beck, that shows that $m(n) \ge \Omega(n^{(1/3)-o(1)}2^n)$.

The best known upper bound to $m(n)$ is found by turning the probabilistic argument "on its head." Basically, the sets become random and each coloring defines an event. Fix V with v points, where we shall later optimize v. Let χ be a coloring of V with a points in one color, $b = v - a$ points in the other. Let $S \subset V$ be a uniformly selected n-set. Then

$$\Pr(S \text{ is monochromatic under } \chi) = \frac{\binom{a}{n} + \binom{b}{n}}{\binom{v}{n}}.$$

Let us assume ν is even for convenience. As $\binom{y}{n}$ is convex, this expression is minimized when $a = b$. Thus

$$\Pr(S \text{ is monochromatic under } \chi) \geq p,$$

where we set

$$p = \frac{2\binom{\nu/2}{n}}{\binom{\nu}{n}}$$

for notational convenience. Now let $S_1, \ldots, S_m$ be uniformly and independently chosen n-sets, m to be determined. For each coloring χ, let A_χ be the event that none of the S_i are monochromatic. By the independence of the S_i

$$\Pr(A_\chi) \leq (1-p)^m.$$

There are 2^ν colorings, so

$$\Pr\left(\bigcup_\chi A_\chi\right) \leq 2^\nu (1-p)^m.$$

When this quantity is less than 1, there exist $S_1, \ldots, S_m$ so that no A_χ holds, i.e., $S_1, \ldots, S_m$ is not 2-colorable—and hence $m(n) \leq m$.

The asymptotics provide a fairly typical example of those encountered when employing the probabilistic method. We first use the inequality $1 - p \leq e^{-p}$. This is valid for all positive p and the terms are quite close when p is small. When

$$m = \left\lceil \frac{\nu \ln 2}{p} \right\rceil,$$

then $2^\nu(1-p)^m < 2^\nu e^{-pm} \leq 1$ so $m(n) \leq m$. Now we need find ν to minimize ν/p. We may interpret p as the probability of picking n balls of one color from an urn with $\nu/2$ white and $\nu/2$ black balls, sampling without replacement. It is tempting to estimate p by 2^{-n+1}, the probability for sampling with replacement. This approximation would yield $m \sim \nu 2^{n-1}(\ln 2)$. As ν gets smaller, however, the approximation becomes less accurate and, as we wish to minimize m, the trade-off becomes essential. We use a second-order approximation

$$p = \frac{2\binom{\nu/2}{n}}{\binom{\nu}{n}} = 2^{1-n} \prod_{i=0}^{n-1} \frac{\nu - 2i}{\nu - i} \sim 2^{1-n} e^{-n^2/2\nu}$$

as long as $\nu \gg n$, estimating

$$\frac{\nu - 2i}{\nu - i} \sim 1 - \frac{i}{\nu} \sim e^{-i/\nu}.$$

Elementary calculus gives $\nu = n^2/2$ for the optimal value. The evenness of ν may require a change of at most 2, which turns out to be asymptotically negligible. This yields the following result of Erdős (1964).

Theorem 3.2.

$$m(n) < (1+o(1))\frac{e\ln 2}{4}n^2 2^n.$$

Let $\mathcal{F} = \{(A_i, B_i)\}_{i=1}^h$ be a family of pairs of subsets of an arbitrary set. We call $\mathcal{F}$ a *(k,ℓ)-system* if $|A_i| = k$ and $|B_i| = \ell$ for all $1 \le i \le h$, $A_i \cap B_i = \emptyset$, and $A_i \cap B_j \ne \emptyset$ for all $1 \le i, j \le h$. B. Bollobás (1965) proved the following result, which has many interesting extensions and applications.

Theorem 3.3. *If $\mathcal{F} = \{(A_i, B_i)\}_{i=1}^h$ is a (k,ℓ)-system, then $h \le \binom{k+\ell}{k}$.*

Proof. Put $X = \bigcup_{i=1}^h (A_i \cup B_i)$ and consider a random order π of X. For each i, $1 \le i \le h$, let X_i be the event that all the elements of A_i precede all those of B_i in this order. Clearly, $\Pr(X_i) = 1/\binom{k+\ell}{k}$. It is also easy to check that the events X_i are pairwise disjoint. Indeed, assume this is false and let π be an order in which all the elements of A_i precede those of B_i and all the elements of A_j precede those of B_j. Without loss of generality we may assume that the last element of A_i does not appear after the last element of A_j. But in this case, all elements of A_i precede all those of B_j, contradicting the fact that $A_i \cap B_j \ne \emptyset$. Therefore, all the events X_i are pairwise disjoint, as claimed. It follows that $1 \ge \Pr(\bigcup_{i=1}^h X_i) = \sum_{i=1}^h \Pr(X_i) = h \cdot 1/\binom{k+\ell}{k}$, completing the proof. ■

Theorem 3.3 is sharp, as shown by the family $\mathcal{F} = \{(A, X \setminus A) : A \subset X,\ |A| = k\}$, where $X = \{1, 2, \ldots, k+\ell\}$.

4. COMBINATORIAL NUMBER THEORY

A subset A of an Abelian group G is called *sum-free* if $(A + A) \cap A = \emptyset$, i.e., if there are no $a_1, a_2, a_3 \in A$ such that $a_1 + a_2 = a_3$.

Theorem 4.1 (Erdős [1965a]). *Every set $B = \{b_1, \ldots, b_n\}$ of n nonzero integers contains a sum-free subset A of size $|A| > \frac{1}{3}n$.*

Proof. Let $p = 3k + 2$ be a prime, which satisfies $p > 2\max_{1 \le i \le n} |b_i|$ and put $C = \{k+1, k+2, \ldots, 2k+1\}$. Observe that C is a sum-free subset of the cyclic group Z_p and that

$$\frac{|C|}{p-1} = \frac{k+1}{3k+1} > \frac{1}{3}.$$

Let us choose at random an integer x, $1 \le x < p$, according to a uniform distribution on $\{1, 2, \ldots, p-1\}$, and define $d_1, \ldots, d_n$ by $d_i \equiv x b_i \pmod{p}$, $0 \le d_i < p$. Trivially, for every fixed i, $1 \le i \le n$, as x ranges over all numbers $1, 2, \ldots, p-1$, d_i ranges over all nonzero elements of Z_p and hence $\Pr(d_i \in$

$C) = |C|/(p-1) > 1/3$. Therefore the expected number of elements b_i such that $d_i \in C$ is more than $n/3$. Consequently, there is an x, $1 \le x < p$, and a subsequence A of B of cardinality $|A| > n/3$, such that $xa(\bmod p) \in C$ for all $a \in A$. This A is clearly sum-free, since if $a_1 + a_2 = a_3$ for some $a_1, a_2, a_3 \in A$, then $xa_1 + xa_2 \equiv xa_3(\bmod p)$, contradicting the fact that C is a sum-free subset of Z_p. This completes the proof. ■

In Alon and Kleitman (1990), it is shown that the constant 1/3 in Theorem 4.1 cannot be replaced by 12/29 (or any bigger constant). The best possible constant is not known.

5. DISJOINT PAIRS

The probabilistic method is most striking when it is applied to prove theorems whose statement does not seem to suggest at all the need for probability. Most of the examples given in the previous sections are simple instances of such statements. In this section we describe a (slightly) more complicated result, due to Alon and Frankl (1985), which solves a conjecture of Daykin and Erdős.

Let $\mathcal{F}$ be a family of m distinct subsets of $X = \{1, 2, \ldots, n\}$. Let $d(\mathcal{F})$ denote the number of disjoint pairs in F, i.e.,

$$d(\mathcal{F}) = |\{(F, F') : F, F' \in \mathcal{F},\ F \cap F' = \emptyset\}|.$$

Daykin and Erdős conjectured that if $m = 2^{(1/2+\delta)n}$, then, for every fixed $\delta > 0$, $d(\mathcal{F}) = o(m^2)$, as n tends to infinity. This result follows from the following theorem, which is a special case of a more general result.

Theorem 5.1. *Let $\mathcal{F}$ be a family of $m = 2^{(1/2+\delta)n}$ subsets of $X = \{1, 2, \ldots, n\}$, where $\delta > 0$. Then*

$$d(\mathcal{F}) < m^{2-\delta^2/2}. \tag{1}$$

Proof. Suppose (1) is false and pick independently t members $A_1, A_2, \ldots, A_t$ of $\mathcal{F}$ with repetitions at random, where t is a large positive integer, to be chosen later. We will show that with positive probability $|A_1 \cup A_2 \cup \cdots \cup A_t| > n/2$ and still this union is disjoint to more than $2^{n/2}$ distinct subsets of X. This contradiction will establish (1). In fact

$$\Pr\left(|A_1 \cup A_2 \cup \cdots \cup A_t| \le \frac{n}{2}\right)$$

$$\le \sum_{S \subset X, |S| \le n/2} \Pr(A_i \subset S,\ i = 1, \ldots, t) \le 2^n \left(\frac{2^{n/2}}{2^{((1/2)+\delta)n}}\right)^t = 2^{n(1-\delta t)}. \tag{2}$$

Define

$$v(B) = |\{A \in \mathcal{F} : B \cap A = \emptyset\}|.$$

Clearly,

$$\sum_{B \in \mathcal{F}} v(B) = 2d(\mathcal{F}) \geq 2m^{2-\delta^2/2}.$$

Let Y be a random variable whose value is the number of members $B \in \mathcal{F}$ that are disjoint to all the A_i $(1 \leq i \leq t)$. By the convexity of z^t, the expected value of Y satisfies

$$E(Y) = \sum_{B \in \mathcal{F}} \left(\frac{v(B)}{m}\right)^t = \frac{1}{m^t} \cdot m \left(\frac{\sum v(B)^t}{m}\right)$$

$$\geq \frac{1}{m^t} \cdot m \left(\frac{2d(\mathcal{F})}{m}\right)^t > 2m^{1-t\delta^2/2}.$$

Since $Y \leq m$, we conclude that

$$\Pr(Y \geq m^{1-t\delta^2/2}) \geq m^{-t\delta^2/2}. \qquad (3)$$

One can check that for $t = \lfloor 1 + 1/(\delta - \delta^2/4 - \delta^3/2) \rfloor$, $m^{1-t\delta^2/2} > 2^{n/2}$ and the right-hand side of (3) is greater than the right-hand side of (2). Thus, with positive probability, $|A_1 \cup A_2 \cup \cdots \cup A_t| > n/2$ and still this union is disjoint to more than $2^{n/2}$ members of F. This contradiction implies inequality (1). ■

The Probabilistic Lens: The Erdős–Ko–Rado Theorem

A family $\mathcal{F}$ of sets is called intersecting if $A, B \in \mathcal{F}$ implies $A \cap B \neq \emptyset$. Suppose $n \geq 2k$ and let $\mathcal{F}$ be an intersecting family of k-element subsets of an n-set, for definiteness $\{0, \ldots, n-1\}$. The Erdős–Ko–Rado theorem is that $|\mathcal{F}| \leq \binom{n-1}{k-1}$. This is achievable by taking the family of k-sets containing a particular point. We give a short proof due to G. Katona (1972).

Lemma. *For* $0 \leq s \leq n-1$, *set* $A_s = \{s, s+1, \ldots, s+k-1\}$ *where addition is modulo n. Then* $\mathcal{F}$ *can contain at most k of the sets* A_s.

Proof. Suppose $A_\ell \in \mathcal{F}$. There are precisely $2k-2$ sets of the form A_s (besides A_ℓ) that intersect A_ℓ, and these can be arranged in $k-1$ pairs of nonintersecting sets. Clearly $\mathcal{F}$ can contain at most one member of each of these pairs, completing the proof of the lemma.

Now we prove the Erdős–Ko–Rado theorem. Let a permutation σ of $\{0, \ldots, n-1\}$ and $i \in \{0, \ldots, n-1\}$ be chosen randomly, uniformly, and independently and set $A = \{\sigma(i), \sigma(i+1), \ldots, \sigma(i+k-1)\}$, addition again modulo n. Conditioning on any choice of σ, the lemma gives $\Pr[A \in \mathcal{F}] \leq k/n$. Hence $\Pr[A \in \mathcal{F}] \leq k/n$. But A is uniformly chosen from all k-sets, so

$$\frac{k}{n} \geq \Pr[A \in \mathcal{F}] = \frac{|\mathcal{F}|}{\binom{n}{k}}$$

and

$$|\mathcal{F}| \leq \frac{k}{n}\binom{n}{k} = \binom{n-1}{k-1}. \qquad \blacksquare$$

2

Linearity of Expectation

1. BASICS

Let $X_1, \ldots, X_n$ be random variables, $X = c_1X_1 + \cdots + c_nX_n$. Linearity of expectation states that

$$E[X] = c_1E[X_1] + \cdots + c_nE[X_n].$$

The power of this principle comes from there being no restrictions on the dependence or independence of the X_i. In many instances, $E[X]$ can be easily calculated by a judicious decomposition into simple (often indicator) random variables X_i.

Let σ be a random permutation on $\{1, \ldots, n\}$, uniformly chosen. Let $X(\sigma)$ be the number of fixed points of σ. To find $E[X]$, we decompose $X = X_1 + \cdots + X_n$, where X_i is the indicator random variable of the event $\sigma(i) = i$. Then

$$E[X_i] = \Pr[\sigma(i) = i] = \frac{1}{n}$$

so that

$$E[X] = \frac{1}{n} + \cdots + \frac{1}{n} = 1.$$

In application we often use the fact that there is a point in the probability space for which $X \geq E[X]$ and a point for which $X \leq E[X]$. We have selected results with a purpose of describing this basic methodology. The following result of Szele (1943) is oftimes considered the first use of the probabilistic method.

Theorem 1.1. *There is a tournament T with n players and at least $n!2^{-(n-1)}$ Hamiltonian paths.*

Proof. In the random tournament, let X be the number of Hamiltonian paths. For each permutation σ, let X_σ be the indicator random variable for σ giving a Hamiltonian path—that is, satisfying $(\sigma(i), \sigma(i+1)) \in T$ for $1 \leq i < n$. Then $X = \sum X_\sigma$ and

$$E[X] = \sum E[X_\sigma] = n!2^{-(n-1)}.$$

Thus some tournament has at least $E[X]$ Hamiltonian paths. ∎

Szele conjectured that the maximum possible number of Hamiltonian paths in a tournament on n players is at most $n!/(2-o(1))^n$. This was proved in Alon (1990b) and is presented in "The Probabilistic Lens: Hamiltonian Paths" (following Chapter 4).

2. SPLITTING GRAPHS

Theorem 2.1. *Let $G=(V,E)$ be a graph with n vertices and e edges. Then G contains a bipartite subgraph with at least $e/2$ edges.*

Proof. Let $T \subseteq V$ be a random subset given by $\Pr[x \in T] = 1/2$, these probabilities mutually independent. Set $B = V - T$. Call an edge $\{x,y\}$ crossing if exactly one of x,y are in T. Let X be the number of crossing edges. We decompose

$$X = \sum_{\{x,y\} \in E} X_{xy},$$

where X_{xy} is the indicator random variable for $\{x,y\}$ being crossing. Then

$$E[X_{xy}] = \tfrac{1}{2}$$

as two fair coin flips have probability $1/2$ of being different. Then

$$E[X] = \sum_{\{x,y\} \in E} E[X_{xy}] = \frac{e}{2}.$$

Thus $X \geq e/2$ for some choice of T, and the set of those crossing edges form a bipartite graph. ■

A more subtle probability space gives a small improvement.

Theorem 2.2. *If G has $2n$ vertices and e edges, then it contains a bipartite subgraph with at least $en/(2n-1)$ edges. If G has $2n+1$ vertices and e edges, then it contains a bipartite subgraph with at least $e(n+1)/(2n+1)$ edges.*

Proof. When G has $2n$ vertices, let T be chosen uniformly from among all n-element subsets of V. Any edge $\{x,y\}$ now has probability $n/(2n-1)$ of being crossing and the proof concludes as before. When G has $2n+1$ vertices, choose T uniformly from among all n-element subsets of V and the proof is similar. ■

Here is a more complicated example in which the choice of distribution requires a preliminary lemma. Let $V = V_1 \cup \cdots \cup V_k$ where the V_i are disjoint sets of size n. Let $h : [V]^k \to \{-1,+1\}$ be a 2-coloring of the k-sets. A k-set E is crossing if it contains precisely one point from each V_i. For $S \subseteq V$, set $h(S) = \sum h(E)$, the sum over all k-sets $E \subseteq S$.

Theorem 2.3. *Suppose $h(E) = +1$ for all crossing k-sets E. Then there is an $S \subseteq V$ for which*

$$|h(S)| \geq c_k n^k.$$

Here c_k is a positive constant, independent of n.

Lemma 2.4. *Let P_k denote the set of all homogeneous polynomials $f(p_1,\ldots,p_k)$ of degree k with all coefficients having absolute value at most 1 and $p_1 p_2 \ldots p_k$ having coefficient 1. Then for all $f \in P_k$, there exist $p_1,\ldots,p_k \in [0,1]$ with*

$$|f(p_1,\ldots,p_k)| \geq c_k.$$

Here c_k is positive and independent of f.

Proof. Set

$$M(f) = \max_{p_1,\ldots,p_k \in [0,1]} |f(p_1,\ldots,p_k)|.$$

For $f \in P_k$, $M(f) > 0$ as f is not the zero polynomial. As P_k is compact and $M : P_k \to R$ is continuous, M must assume its minimum c_k. ■

Proof of Theorem 2.3. Define a random $S \subseteq V$ by setting

$$\Pr[x \in S] = p_i, \qquad x \in V_i,$$

these probabilities mutually independent, p_i to be determined. Set $X = h(S)$. For each k-set E, set

$$X_E = \begin{cases} h(E) & \text{if } E \subseteq S \\ 0 & \text{otherwise.} \end{cases}$$

Say E has type $(a_1,\ldots,a_k)$ if $|E \cap V_i| = a_i$, $1 \leq i \leq k$. For these E,

$$E[X_E] = h(E)\Pr[E \subseteq S] = h(E)p_1^{a_1} \cdots p_k^{a_k}.$$

Combining terms by type,

$$E[X] = \sum_{a_1+\cdots+a_k=k} p_1^{a_1} \cdots p_k^{a_k} \sum_{E \text{ of type } (a_1,\ldots,a_k)} h(E).$$

When $a_1 = \cdots = a_k = 1$, all $h(E) = 1$ by assumption, so

$$\sum_{E \text{ of type } (1,\ldots,1)} h(E) = n^k.$$

For any other type, there are fewer than n^k terms, each ± 1, so

$$\sum_{E \text{ of type } (a_1,\ldots,a_k)} h(E) \leq n^k.$$

Thus

$$E[X] = n^k f(p_1, \ldots, p_k)$$

where $f \in P_k$, as defined by Lemma 2.4.

Now select $p_1, \ldots, p_k \in [0,1]$ with $|f(p_1, \ldots, p_k)| \geq c_k$. Then

$$E[|X|] \geq |E[X]| \geq c_k n^k.$$

Some particular value of $|X|$ must exceed or equal its expectation. Hence there is a particular set $S \subseteq V$ with

$$|X| = |h(S)| \geq c_k n^k. \qquad \blacksquare$$

Theorem 2.3 has an interesting application to Ramsey theory. It is known (see Erdős [1965b]) that given any coloring with 2 colors of the k-sets of an n-set, there exist k disjoint m-sets, $m = \Theta((\ln n)^{1/(k-1)})$, so that all crossing k-sets are the same color. From Theorem 2.3 there then exists a set of size $\Theta((\ln n)^{1/(k-1)})$, at least $1/2 + \epsilon_k$ of whose k-sets are the same color. This is somewhat surprising, since it is known that there are colorings in which the largest monochromatic set has size at most the $k-2$-fold logarithm of n.

3. TWO QUICKIES

Linearity of expectation sometimes gives very quick results.

Theorem 3.1. *There is a 2-coloring of K_n with at most*

$$\binom{n}{a} 2^{1-\binom{a}{2}}$$

monochromatic K_a.

Proof (Outline). Take a random coloring. Let X be the number of monochromatic K_a and find $E[X]$. For some coloring, the value of X is at most this expectation. $\blacksquare$

In Chapter 15 it is shown how such a coloring can be found deterministically and efficiently.

Theorem 3.2. *There is a 2-coloring of $K_{m,n}$ with at most*

$$\binom{m}{a}\binom{n}{b} 2^{1-ab} + \binom{n}{a}\binom{m}{b} 2^{1-ab}$$

monochromatic $K_{a,b}$.

Proof (Outline). Take a random coloring. Let X be the number of monochromatic $K_{a,b}$ and find $E[X]$. For some coloring, the value of X is at most this expectation. ■

4. BALANCING VECTORS

The next result has an elegant *non*probabilistic proof, which we defer to the end of this chapter. Here $|v|$ is the usual Euclidean norm.

Theorem 4.1. *Let $v_1,\ldots,v_n \in R^n$, all $|v_i| = 1$. Then there exist $\epsilon_1,\ldots,\epsilon_n = \pm 1$ so that*

$$|\epsilon_1 v_1 + \cdots + \epsilon_n v_n| \le \sqrt{n},$$

and also there exist $\epsilon_1,\ldots,\epsilon_n = \pm 1$ so that

$$|\epsilon_1 v_1 + \cdots + \epsilon_n v_n| \ge \sqrt{n}.$$

Proof. Let $\epsilon_1,\ldots,\epsilon_n$ be selected uniformly and independently from $\{-1, +1\}$. Set

$$X = |\epsilon_1 v_1 + \cdots + \epsilon_n v_n|^2.$$

Then

$$X = \sum_{i=1}^{n} \sum_{j=1}^{n} \epsilon_i \epsilon_j v_i \cdot v_j.$$

Thus

$$E[X] = \sum_{i=1}^{n} \sum_{j=1}^{n} v_i \cdot v_j E[\epsilon_i \epsilon_j].$$

When $i \ne j$, $E[\epsilon_i \epsilon_j] = E[\epsilon_i]E[\epsilon_j] = 0$. When $i = j$, $\epsilon_i^2 = 1$ so $E[\epsilon_i^2] = 1$. Thus

$$E[X] = \sum_{i=1}^{n} v_i \cdot v_i = n.$$

Hence there exist specific $\epsilon_1,\ldots,\epsilon_n = \pm 1$ with $X \ge n$ and with $X \le n$. Taking square roots gives the theorem. ■

The next result includes part of Theorem 4.1 as a linear translate of the $p_1 = \cdots = p_n = 1/2$ case.

Theorem 4.2. *Let $v_1,\ldots,v_n \in R^n$, all $|v_i| \le 1$. Let $p_1,\ldots,p_n \in [0,1]$ be arbitrary and set $w = p_1 v_1 + \cdots + p_n v_n$. Then there exist $\epsilon_1,\ldots,\epsilon_n \in \{0,1\}$ so that, setting $v = \epsilon_1 v_1 + \cdots + \epsilon_n v_n$,*

$$|w - v| \le \frac{\sqrt{n}}{2}.$$

Proof. Pick ϵ_i independently with

$$\Pr[\epsilon_i = 1] = p_i, \qquad \Pr[\epsilon_i = 0] = 1 - p_i.$$

The random choice of ϵ_i gives a random v and a random variable

$$X = |w - v|^2.$$

We expand

$$X = \left|\sum_{i=1}^{n}(p_i - \epsilon_i)v_i\right|^2 = \sum_{i=1}^{n}\sum_{j=1}^{n} v_i \cdot v_j(p_i - \epsilon_i)(p_j - \epsilon_j)$$

so that

$$E[X] = \sum_{i=1}^{n}\sum_{j=1}^{n} v_i \cdot v_j E[(p_i - \epsilon_i)(p_j - \epsilon_j)].$$

For $i \neq j$,

$$E[(p_i - \epsilon_i)(p_j - \epsilon_j)] = E[p_i - \epsilon_i]E[p_j - \epsilon_j] = 0.$$

For $i = j$,

$$E[(p_i - \epsilon_i)^2] = p_i(p_i - 1)^2 + (1 - p_i)p_i^2 = p_i(1 - p_i) \leq \tfrac{1}{4}.$$

($E[(p_i - \epsilon_i)^2] = \mathrm{var}[\epsilon_i]$, the *variance* to be discussed in Chapter 4.) Thus

$$E[X] = \sum_{i=1}^{n} p_i(1 - p_i)|v_i|^2 \leq \frac{1}{4}\sum_{i=1}^{n}|v_i|^2 = \frac{n}{4}$$

and the proof concludes as in that of Theorem 4.1. ■

5. UNBALANCING LIGHTS

Theorem 5.1. *Let $a_{ij} = \pm 1$ for $1 \leq i, j \leq n$. Then there exist $x_i, y_j = \pm 1$, $1 \leq i, j \leq n$ so that*

$$\sum_{i=1}^{n}\sum_{j=1}^{n} a_{ij}x_i y_j \geq \left(\sqrt{\frac{2}{\pi}} + o(1)\right) n^{3/2}.$$

This result has an amusing interpretation. Let an $n \times n$ array of lights be given, each either on ($a_{ij} = +1$) or off ($a_{ij} = -1$). Suppose for each row and each column there is a switch so that if the switch is pulled ($x_i = -1$ for row i and $y_j = -1$ for column j), all of the lights in that line are "switched": on to off or off to on. Then for any initial configuration it is possible to perform switches so that the number of lights on minus the number of lights off is at least $(\sqrt{2/\pi} + o(1))n^{3/2}$.

Proof of Theorem 5.1. Forget the x's. Let $y_1,\ldots,y_n = \pm 1$ be selected independently and uniformly and set

$$R_i = \sum_{j=1}^{n} a_{ij} y_j,$$

$$R = \sum_{i=1}^{n} |R_i|.$$

Fix i. Regardless of a_{ij}, $a_{ij}y_j$ is $+1$ or -1 with probability $1/2$ and their values (over j) are independent. (That is, whatever the ith row is initially, after random switching it becomes a uniformly distributed row, all 2^n possibilities equally likely.) Thus R_i has distribution S_n—the distribution of the sum of n independent uniform $\{-1, 1\}$ random variables—and so

$$E[|R_i|] = E[|S_n|] = \left(\sqrt{\frac{2}{\pi}} + o(1)\right)\sqrt{n}.$$

These asymptotics may be found by estimating S_n by $\sqrt{n}N$, where N is standard normal and using elementary calculus. Alternatively, a closed form

$$E[|S_n|] = n2^{1-n}\binom{n-1}{\lfloor (n-1)/2 \rfloor}$$

may be derived combinatorially (a problem in the 1974 Putnam competition!) and the asymptotics follows from Stirling's formula.

Now apply linearity of expectation to R:

$$E[R] = \sum_{i=1}^{n} E[|R_i|] = \left(\sqrt{\frac{2}{\pi}} + o(1)\right) n^{3/2}.$$

There exist $y_1,\ldots,y_n = \pm 1$ with R at least this value. Finally, pick x_i with the same sign as R_i, so that

$$\sum_{i=1}^{n} x_i \sum_{j=1}^{n} a_{ij} y_j = \sum_{i=1}^{n} x_i R_i = \sum_{i=1}^{n} |R_i| = R \geq \left(\sqrt{\frac{2}{\pi}} + o(1)\right) n^{3/2}. \qquad \blacksquare$$

Another result on unbalancing lights appears in "The Probabilistic Lens: Unbalancing Lights" (after Chapter 12).

6. WITHOUT COIN FLIPS

A nonprobabilistic proof of Theorem 2.1 may be given by placing each vertex in either T or B sequentially. At each stage, place x in either T or B so that at least half of the edges from x to previous vertices are crossing. With this effective algorithm, at least half the edges will be crossing.

There is also a simple sequential algorithm for choosing signs in Theorem 4.1. When the sign for v_i is to be chosen, a partial sum $w = \epsilon_1 v_1 + \cdots + \epsilon_{i-1}v_{i-1}$ has been calculated. Now if it is desired that the sum be small, select $\epsilon_i = \pm 1$ so that $\epsilon_i v_i$ makes an acute (or right) angle with w. If the sum need be big, make the angle obtuse or right. In the extreme case when all angles are right angles, Pythagoras and induction give that the final w has norm $\sqrt{n}$, otherwise it is either less than $\sqrt{n}$ or greater than $\sqrt{n}$ as desired.

For Theorem 4.2, a greedy algorithm produces the desired ϵ_i. Given $v_1,\dots,v_n \in R^n$, $p_1,\dots,p_n \in [0,1]$, suppose $\epsilon_1,\dots,\epsilon_{s-1} \in \{0,1\}$ have already been chosen. Set $w_{s-1} = \sum_{i=1}^{s-1}(p_i - \epsilon_i)v_i$, the partial sum. Select ϵ_s so that

$$w_s = w_{s-1} + (p_s - \epsilon_s)v_s = \sum_{i=1}^{s}(p_i - \epsilon_i)v_i$$

has minimal norm. A random $\epsilon_s \in \{0,1\}$ chosen with $\Pr[\epsilon_s = 1] = p_s$ gives

$$\begin{aligned} E[|w_s|^2] &= |w_{s-1}|^2 + 2w_{s-1}\cdot v_s E[p_s - \epsilon_s] + |v_s|^2 E(p_s - \epsilon_s)^2 \\ &= |w_{s-1}|^2 + p_s(1-p_s)|v_s|^2, \end{aligned}$$

so for some choice of $\epsilon_s \in \{0,1\}$,

$$|w_s|^2 \le |w_{s-1}|^2 + p_s(1-p_s)|v_s|^2.$$

As this holds for all $1 \le s \le n$ (taking $w_0 = 0$), the final

$$|w_n|^2 \le \sum_{i=1}^{n} p_i(1-p_i)|v_i|^2.$$

While the proofs appear similar, a direct implementation of the proof of Theorem 4.2 to find $\epsilon_1,\dots,\epsilon_n$ might take an exhaustive search with exponential time. In applying the greedy algorithm at the sth stage, one makes two calculations of $|w_s|^2$, depending on whether $\epsilon_s = 0$ or 1, and picks that ϵ_s giving the smaller value. Hence there are only a linear number of calculations of norms to be made and the entire algorithm takes only quadratic time. In Chapter 15 we discuss several similar examples in a more general setting.

The Probabilistic Lens: Brégman's Theorem

Let $A = [a_{ij}]$ be an $n \times n$ matrix with all $a_{ij} \in \{0,1\}$. Let $r_i = \sum_{1 \le j \le n} a_{ij}$ be the number of ones in the ith row. Let S be the set of permutations $\sigma \in S_n$ with $a_{i,\sigma i} = 1$ for $1 \le i \le n$. Then the permanent per(A) is simply $|S|$. The following result was conjectured by Minc and proved by Brégman (1973). The proof presented here is similar to that of Schrijver (1978).

Brégman's Theorem.

$$\text{per}(A) \le \prod_{1 \le i \le n} (r_i!)^{1/r_i}.$$

Pick $\sigma \in S$ and $\tau \in S_n$ independently and uniformly. Set $A^1 = A$. Let $R_{\tau 1}$ be the number of ones in row $\tau 1$ in A^1. Delete row $\tau 1$ and column $\sigma\tau 1$ from A^1 to give A^2. In general, let A^i denote A with rows $\tau 1, \ldots, \tau(i-1)$ and columns $\sigma 1, \ldots, \sigma\tau(i-1)$ deleted and let $R_{\tau i}$ denote the number of ones of row τi in A^i. (This is nonzero as the $\sigma\tau i$th column has a one.) Set

$$L = L(\sigma, \tau) = \prod_{1 \le i \le n} R_{\tau i}.$$

We think, roughly, of L as Lazyman's permanent calculation. There are $R_{\tau 1}$ choices for a one in row $\tau 1$, each of which leads to a different subpermanent calculation. Instead, Lazyman takes the factor $R_{\tau 1}$, takes the one from permutation σ, and examines A^2. As $\sigma \in S$ is chosen uniformly, Lazyman tends toward the high subpermanents and so it should not be surprising that he tends to overestimate the permanent. To make this precise, we define the geometric mean $G[Y]$. If $Y > 0$ takes values $a_1, \ldots, a_s$ with probabilities $p_1, \ldots, p_s$, respectively, then $G[Y] = \prod a_i^{p_i}$. Equivalently, $G[Y] = e^{E[\ln Y]}$. Linearity of expectation translates into the geometric mean of a product being the product of the geometric means.

Claim.

$$\text{per}(A) \le G[L].$$

We show this for any fixed τ. Set $\tau 1 = 1$ for convenience of notation. We use induction on the size of the matrix. Reorder, for convenience, so that the first row has ones in the first r columns where $r = r_1$. For $1 \le j \le r$, let t_j be the permanent of A with the first row and jth column removed or, equivalently, the number of $\sigma \in S$ with $\sigma 1 = j$. Set

$$t = \frac{t_1 + \cdots + t_r}{r}$$

so that $\mathrm{per}(A) = rt$. Conditioning on $\sigma 1 = j$, $R_2 \ldots R_n$ is Lazyman's calculation of $\mathrm{per}(A^2)$, where A^2 is A with the first row and jth column removed. By induction,

$$G[R_2 \ldots R_n \mid \sigma 1 = j] \ge t_j,$$

and so

$$G[L] \ge \prod_{j=1}^{r} (rt_j)^{t_j/\mathrm{per}(A)} = r \prod_{j=1}^{r} t_j^{t_j/rt}.$$

Lemma.

$$\left(\prod_{j=1}^{r} t_j^{t_j} \right)^{1/r} \ge t^t.$$

Proof. Taking logarithms, this is equivalent to

$$\frac{1}{r} \sum_{j=1}^{r} t_j \ln t_j \ge t \ln t,$$

which follows from the convexity of the function $f(x) = x \ln x$.

Applying the lemma,

$$G[L] \ge r \prod_{j=1}^{r} t_j^{t_j/rt} \ge r(t^t)^{1/t} = rt = \mathrm{per}(A). \quad \blacksquare$$

Now we calculate $G[L]$ conditional on a fixed σ. For convenience of notation, reorder so that $\sigma i = i$, all i, and assume that the first row has ones in precisely the first r_1 columns. With τ selected uniformly, the columns $1, \ldots, r_1$ are deleted in order uniform over all $r_1!$ possibilities. R_1 is the number of those columns remaining when the first column is to be deleted. As the first column is equally likely to be in any position among those r_1 columns, R_1 is uniformly distributed from 1 to r_1 and $G[R_1] = (r_1!)^{1/r_1}$. "Linearity" then gives

$$G[L] = G\left[\prod_{i=1}^{n} R_i \right] = \prod_{i=1}^{n} G[R_i] = \prod_{i=1}^{n} (r_i!)^{1/r_i}.$$

The overall $G[L]$ is the geometric mean of the conditional $G[L]$ and thus has the same value. That is,

$$\text{per}(A) \leq G[L] = \prod_{i=1}^{n} (r_i!)^{1/r_i}.$$

3

Alterations

The basic probabilistic method was described in Chapter 1 as follows: Trying to prove that a structure with certain desired properties exists, one defines an appropriate probability space of structures and then shows that the desired properties hold in this space with positive probabilities. In this chapter, we consider situations where the "random" structure does not have all the desired properties but may have a few "blemishes." With a small alteration, we remove the blemishes, giving the desired structure.

1. RAMSEY NUMBERS

Recall from Section 1 of Chapter 1 that $R(k,l) > n$ means there exists a 2-coloring of the edges of K_n by red and blue so that there is neither a red K_k nor a blue K_l.

Theorem 1.1. *For any integer n*

$$R(k,k) > n - \binom{n}{k} 2^{1-\binom{k}{2}}.$$

Proof. Consider a random 2-coloring of the edges of K_n obtained by coloring each edge independently either red or blue, where each color is equally likely. For any set R of k vertices, let X_R be the indicator random variable for the event that the induced subgraph of K_n on R is monochromatic. Set $X = \sum X_R$, the sum over all such R. From linearity of expectation,

$$E[X] = \sum E[X_R] = m \qquad \text{with} \quad m = \binom{n}{k} 2^{1-\binom{k}{2}}.$$

Thus there exists a 2-coloring for which $X \leq m$. Fix such a coloring. Remove from K_n one vertex from each monochromatic k-set. At most m vertices have been removed (we may have "removed" the same vertex more than once but this only helps), so s vertices remain with $s \geq n - m$. This coloring on these s points has no monochromatic k-set. ∎

We are left with the "calculus" problem of finding that n that will optimize the inequality. Some analysis shows that we should take $n \sim e^{-1}k2^{k/2}(1 - o(1))$, giving

$$R(k,k) > \frac{1}{e}(1 + o(1))k2^{k/2}.$$

A careful examination of Proposition 1.1 proved in Chapter 1 gives the lower bound

$$R(k,k) > \frac{1}{e\sqrt{2}}(1 + o(1))k2^{k/2}.$$

The more powerful Lovász local lemma—see Chapter 5—gives

$$R(k,k) > \frac{\sqrt{2}}{e}(1 + o(1))k2^{k/2}.$$

The distinctions between these bounds may be considered inconsequential, since the best known upper bound for $R(k,k)$ is on the order $(4 + o(1))^n$. The upper bounds do not involve probabilistic methods and may be found, for example, in Graham, Rothschild, and Spencer (1990). We give all three lower bounds in following our philosophy of emphasizing *methodologies* rather than results.

In dealing with the off-diagonal Ramsey numbers, the distinction between the basic method and the alteration is given in the following two results.

Theorem 1.2. *If there exists $p \in [0,1]$ with*

$$\binom{n}{k} p^{\binom{k}{2}} + \binom{n}{l} (1-p)^{\binom{l}{2}} < 1,$$

then $R(k,l) > n$.

Theorem 1.3. *For all integers n and $p \in [0,1]$,*

$$R(k,l) > n - \binom{n}{k} p^{\binom{k}{2}} - \binom{n}{l} (1-p)^{\binom{l}{2}}.$$

Proofs. In both cases, we consider a random 2-coloring of K_n obtained by coloring each edge independently either red or blue, where each edge is red with probability p. Let X be the number of red k-sets plus the number of blue l-sets. Linearity of expectation gives

$$E[X] = \binom{n}{k} p^{\binom{k}{2}} + \binom{n}{l} (1-p)^{\binom{l}{2}}.$$

For Theorem 1.2, $E[X] < 1$, so there exists a 2-coloring with $X = 0$. For Theorem 1.3, there exists a 2-coloring with s "bad" sets (either red k-sets or blue l-sets), $s \leq E[X]$. Removing one point from each bad set gives a coloring of at least $n - s$ points with no bad sets. ∎

The asymptotics of Theorems 1.2 and 1.3 can get fairly complex. Often Theorem 1.3 gives a substantial improvement over Theorem 1.2. Even further improvements may be found using the Lovász local lemma. These bounds have been analyzed in Spencer (1977).

2. INDEPENDENT SETS

Here is a short and sweet argument that gives roughly half of the celebrated Turán's theorem. $\alpha(G)$ is the independence number of a graph G, $\alpha(G) \geq t$ means there exist t vertices with no edges between them.

Theorem 2.1. *Let $G = (V,E)$ have v vertices and $nd/2$ edges, $d \geq 1$. Then $\alpha(G) \geq n/2d$.*

Proof. Let $S \subseteq V$ be a random subset defined by

$$\Pr[v \in S] = p,$$

p to be determined, the events $v \in S$ being mutually independent. Let $X = |S|$ and let Y be the number of edges in $G|_S$. For each $e = \{i,j\} \in E$, let Y_e be the indicator random variable for the event $i,j \in S$ so that $Y = \sum_{e \in E} Y_e$. For any such e

$$E[Y_e] = \Pr[i,j \in S] = p^2,$$

so by linearity of expectation,

$$E[Y] = \sum_{e \in E} E[Y_e] = \frac{nd}{2}p^2.$$

Clearly, $E[X] = np$, so, again by linearity of expectation,

$$E[X - Y] = np - \frac{nd}{2}p^2.$$

We set $p = 1/d$ (here using $d \geq 1$) to maximize this quantity, giving

$$E[X - Y] = \frac{n}{2d}.$$

Thus there exists a specific S for whom the number of vertices of S minus the number of edges in S is at least $n/2d$. Select one vertex from each edge of S and delete it. This leaves a set S^* with at least $n/2d$ vertices. All edges having been destroyed, S^* is an independent set. ∎

The full result of Turán is given in "The Probabilistic Lens: Turán's Theorem" (following Chapter 6).

3. COMBINATORIAL GEOMETRY

For a set S of n points in the unit square U, let $T(S)$ be the minimum area of a triangle whose vertices are three distinct points of S. Put $T(n) = \max T(S)$, where S ranges over all sets of n points in U. Heilbronn conjectured that $T(n) = O(1/n^2)$. This conjecture was disproved by Komlós, Pintz, and Szemerédi (1982), who showed, by a rather involved probabilistic construction, that there is a set S of n points in U such that $T(S) = \Omega(\log n/n^2)$. As this argument is rather complicated, we only present here a simpler one showing that $T(n) = \Omega(1/n^2)$.

Theorem 3.1. *There is a set S of n points in the unit square U such that $T(S) \geq 1/(100n^2)$.*

Proof. We first make a calculation. Let P, Q, R be independently and uniformly selected from U and let $\mu = \mu(PQR)$ denote the area of the triangle PQR. We bound $\Pr[\mu \leq \epsilon]$ as follows. Let x be the distance from P to Q so that $\Pr[b \leq x \leq b + \Delta b] \leq \pi(b + \Delta b)^2 - \pi b^2$ and in the limit $\Pr[b \leq x \leq b + db] \leq 2\pi b db$. Given P, Q at distance b, the altitude from R to the line PQ must have height $h \leq 2\epsilon/b$ and so R must lie in a strip of width $4\epsilon/b$ and length at most $\sqrt{2}$. This occurs with probability at most $4\sqrt{2}\epsilon/b$. As $0 \leq b \leq \sqrt{2}$, the total probability is bounded by

$$\int_0^{\sqrt{2}} (2\pi b)\left(\frac{4\sqrt{2}\epsilon}{b}\right) db = 16\pi\epsilon.$$

Now let $P_1, \ldots, P_{2n}$ be selected uniformly and independently in U and let X denote the number of triangles $P_iP_jP_k$ with area less than $1/(100n^2)$. For each particular i, j, k, the probability of this occurring is less than $0.6n^{-2}$ and so

$$E[X] \leq \binom{2n}{3}(0.6n^{-2}) < n.$$

Therefore there exists a specific set of $2n$ vertices with fewer than n triangles of area less than $1/(100n^2)$. Delete one vertex from the set from each such triangle. This leaves at least n vertices and now no triangle has area less than $1/(100n^2)$. ■

We note the following construction of Erdős, showing $T(n) \geq 1/(2(n-1)^2)$ with n prime. On $[0, n-1] \times [0, n-1]$, consider the n points (x, x^2) where x^2 is reduced mod n. (More formally, (x, y) where $y \equiv x^2 \bmod n$ and $0 \leq y < n$.) If some three points of this set were collinear, they would line on a line $y = mx + b$ and m would be a rational number with denominator less than n. But then in Z_n^2, the parabola $y = x^2$ would intersect the line $y = mx + b$ in three points, so that the quadratic $x^2 - mx - b$ would have three distinct roots, an impossibility. Triangles between lattice points in the plane have as their areas either half-integers or integers, hence the areas must be at least $1/2$. Contracting the plane by an $n-1$ factor in both coordinates gives the desired set. While this gem does better than Theorem 3.1, it does not lead to the improvements of Komlós, Pintz, and Szemerédi.

4. PACKING

Let C be a bounded measurable subset of R^d and let $B(x)$ denote the cube $[0,x]^d$ of side x. A *packing* of C into $B(x)$ is a family of mutually disjoint copies of C, all lying inside $B(x)$. Let $f(x)$ denote the largest size of such a family. The packing constant $\delta = \delta(C)$ is defined by

$$\delta(C) = \frac{1}{\mu(C)} \lim_{x\to\infty} f(x)x^{-d},$$

the maximal proportion of space that may be packed by copies of C. (This limit can be proven always to exist, but even without that result the following result holds with lim replaced by lim inf.)

Theorem 4.1. *Let C be bounded, convex, and centrally symmetric around the origin. Then*

$$\delta(C) \geq 2^{-d-1}.$$

Proof. Let P, Q be selected independently and uniformly from $B(x)$ and consider the event $(C+P)\cap(C+Q) \neq \emptyset$. For this to occur, we must have, for some $c_1, c_2 \in C$,

$$P - Q = c_1 - c_2 = 2\frac{c_1 - c_2}{2} \in 2C$$

by central symmetry and convexity. The event $P \in Q + 2C$ has probability at most $\mu(2C)x^{-d}$ for each given Q, hence

$$\Pr[(C+P)\cap(C+Q) \neq \emptyset] \leq \mu(2C)x^{-d} = 2^d x^{-d}\mu(C).$$

Now let $P_1, \ldots, P_n$ be selected independently and uniformly from $B(x)$ and let X be the number of $i<j$ with $(C+P_i)\cap(C+P_j)\neq\emptyset$. From linearity of expectation,

$$E[X] \leq \frac{n^2}{2} 2^d x^{-d} \mu(C).$$

Thus there exists a specific choice of n points with fewer than that many intersecting copies of C. For each P_i, P_j with $(C+P_i)\cap(C+P_j)\neq\emptyset$, remove either P_i or P_j from the set. This leaves at least $n-(n^2/2)2^d x^{-d}\mu(C)$ nonintersecting copies of C. Set $n = x^d 2^{-d}/\mu(C)$ to maximize this quantity, so that there are at least $x^d 2^{-d-1}/\mu(C)$ nonintersecting copies of C. These do not all lie inside $B(x)$, but, letting w denote an upper bound on the absolute values of the coordinates of the points of C, they do all lie inside a cube of side $x+2w$. Hence

$$f(x+2w) \geq \frac{x^d 2^{-d-1}}{\mu(C)}$$

and so

$$\delta(C) \geq \lim_{x\to\infty} \frac{f(x+2w)(x+2w)^{-d}}{\mu(C)} \geq 2^{-d-1}. \qquad \blacksquare$$

A simple greedy algorithm does somewhat better. Let $P_1,\ldots,P_m$ be *any* maximal subset of $[0,x]^d$ with the property that the sets $C+P_i$ are disjoint. We have seen that $C+P_i$ overlaps $C+P$ if and only if $P \in 2C+P_i$. Hence the sets $2C+P_i$ must cover $[0,x]^d$. As each such set has measure $\mu(2C) = 2^d\mu(C)$ we must have $m \geq x^d 2^{-d}/\mu(C)$. As before, all sets $C+P_i$ lie in a cube of side $x+2w$, w a constant, so that

$$f(x+2w) \geq m \geq x^d 2^{-d}/\mu(C)$$

and so

$$\delta(C) \geq 2^{-d}.$$

A still further improvement appears in "The Probabilistic Lens: Efficient Packing."

5. RECOLORING

Suppose that a random coloring leaves a set of blemishes. Here we apply a random recoloring to the blemishes to remove them. If the recoloring is too weak, then not all the blemishes are removed. If the recoloring is too strong, then new blemishes are created. We use the notation of Chapter 1, Section 3, on property B: $m(n) > m$ means that given any n-uniform hypergraph $H = (V,E)$ with m edges, there exists a 2-coloring of V so that no edge is monochromatic. J. Beck (1978) proved $m(n) = \Omega(2^n n^{1/3})$. Here we use his method to give a slightly weaker result.

Theorem 5.1. *If there exists p, $0 \leq p \leq 1/2$ and k, with*

$$2ke^{-pn} + k^2 p e^{pn} < 1,$$

then $m(n) > 2^{n-1}k$.

Corollary 5.2.

$$m(n) = \Omega(2^{n-1} n^{1/3} (\ln n)^{-1/2}).$$

Proof of Corollary 5.2. Take $p = (\ln n)/3n$ for $k < cn^{1/3}(\ln n)^{-1/2}$ with $c < \sqrt{3}$. $\blacksquare$

Proof of Theorem 5.1. Fix $H=(V,E)$ with $m=2^{n-1}k$ edges and p satisfying the condition. First color V randomly, each $v \in V$ is red or blue with independent probability .5. Call this the first coloring. Now for each $v \in V$ that lies in a monochromatic $e \in E$, change the color of v with probability p. (Even if v lies in many monochromatic e, the probability of its color changing is still precisely p. The color changes, of course, are done independently.) The new coloring is called the recoloring.

For each $e \in E$, let A_e be the event that e is monochromatic in both the first coloring and the recoloring. Once e is monochromatic in the first coloring, each $v \in e$ will "flip its coin" once again, so that

$$\Pr[A_e] = 2^{1-n}(p^n + (1-p)^n) \le 2^{1-n}(2(1-p)^n)) \le 2^{1-n}(2e^{-pn})$$

so that

$$\Pr\left[\bigvee A_e\right] \le m2^{1-n}(2e^{-pn}) = 2ke^{-pn}.$$

For each $e,f \in E$ with $e \cap f \neq \emptyset$ let A_{ef} be the event that e was red in the first coloring and f was blue in the recoloring, or the same with red/blue reversed. For every $W \subseteq f-e$, let A_{efW} be the event that $e \cup W$ was red in the first coloring, $f-(e\cap f)-W$ was blue in the first coloring, and f was blue in the recoloring—or the same with red/blue reversed. Then $A_{ef} = \vee A_{efW}$, the disjunction over all $W \subseteq f-e$.

Set $|e \cap f| = t$ and $|W| = w$. Then

$$\Pr[A_{efW}] \le 2 \cdot 2^{-2n+t}p^{t+w}.$$

The factor 2 is the choice of red or blue. The factor 2^{-2n+t} is the probability of achieving the precise first coloring of $e \cup f$. The probability p^{t+w} bounds the probability that every point of $(e\cap f)\cup W$ changes its color in the recoloring. (p^{t+w} is an upper bound, since also every point of W must lie in a set monochromatic under the first coloring and also no other point of f can have its color changed. Attempts to improve Beck's result have centered at this point but have been unsuccessful thus far.) Summing over all W,

$$\Pr[A_{ef}] = \sum_{W\subseteq f-e} \Pr[A_{efW}] \le \sum_{w=0}^{n-t}\binom{n-t}{w}(2^{1-2n+t}p^t)p^w$$

$$= 2^{1-2n+t}p^t(1+p)^{n-t} \le 2^{1-2n}e^{pn}\left(\frac{2p}{1+p}\right)^t.$$

As $p \le 1/2$, $2p/(1+p) < 1$ and the last expression is maximized when $t=1$. Thus

$$\Pr[A_{ef}] \le 2^{1-2n}e^{pn}\frac{2p}{1+p} \le 2^{2-2n}e^{pn}p.$$

There are less than m^2 intersecting pairs $e,f \in E$, so

$$\Pr\left[\bigvee A_{ef}\right] \le m^2 2^{2-2n}e^{pn}p = k^2e^{pn}p$$

and hence

$$\Pr\left[\bigvee A_e \vee \bigvee A_{ef}\right] < 1.$$

There is a first coloring and recoloring for which none of the A_e nor any of the A_{ef} hold. But we have covered all ways a set can be monochromatic under the recoloring—therefore there is a first coloring and recoloring for which there are no monochromatic sets in the recoloring and so $m(n) > m$. ■

The Probabilistic Lens: High Girth and High Chromatic Number

Many consider this one of the most pleasing uses of the probabilistic method, as the result is surprising and does not appear to call for nonconstructive techniques. The *girth* of a graph G is the size of its smallest circuit. $\alpha(G)$ is the size of the largest independent set in G and $\chi(G)$ denotes its chromatic number.

Theorem (Erdős [1959]). *For all k,l there exists a graph G with girth$(G) > l$ and $\chi(G) > k$.*

Proof. Fix $\theta < 1/l$ and let $G \sim G(n,p)$ with $p = n^{\theta-1}$. (That is, G is a random graph on n vertices chosen by picking each pair of vertices as an edge randomly and independently with probability p). Let X be the number of circuits of size at most l. Then

$$E[X] = \sum_{i=3}^{l} \frac{(n)_i}{2i} p^i \leq \sum_{i=3}^{l} \frac{n^{\theta i}}{2i} = o(n)$$

as $\theta l < 1$. In particular,

$$\Pr[X \geq n/2] = o(1).$$

Set $x = \lceil (3/p) \ln n \rceil$ so that

$$\Pr[\alpha(G) \geq x] \leq \binom{n}{x} (1-p)^{\binom{x}{2}} < \left[ne^{-p(x-1)/2}\right]^x = o(1).$$

Let n be sufficiently large so that both these events have probability less than .5. Then there is a specific G with less than $n/2$ cycles of length less than l and with $\alpha(G) < 3n^{1-\theta} \ln n$. Remove from G a vertex from each cycle of length at most l. This gives a graph G^* with at least $n/2$ vertices. G^* has girth greater than l and $\alpha(G^*) \leq \alpha(G)$. Thus

$$\chi(G^*) \geq \frac{|G^*|}{\alpha(G^*)} \geq \frac{n/2}{3n^{1-\theta} \ln n} = \frac{n^\theta}{6 \ln n}.$$

To complete the proof, let n be sufficiently large so that this is greater than k. ■

4

The Second Moment

1. BASICS

After the expectation, the most vital statistic for a random variable X is the *variance*. We denote it var$[X]$. It is defined by

$$\mathrm{var}[X] = E[(X - E[X])^2]$$

and measures how spread out X is from its expectation. We shall generally, following standard practice, let μ denote expectation and σ^2 denote variance. The positive square root σ of the variance is called the *standard deviation*. With this notation, here is our basic tool.

Chebyschev's Inequality. *For any positive* λ,

$$\Pr[|X - \mu| \geq \lambda\sigma] \leq \frac{1}{\lambda^2}.$$

Proof.

$$\sigma^2 = \mathrm{var}[X] = E[(X - \mu)^2] \geq \lambda^2\sigma^2 \Pr[|X - \mu| \geq \lambda\sigma]. \quad \blacksquare$$

The use of Chebyschev's inequality is called the *second moment method*. Chebyschev's inequality is best possible when no additional restrictions are placed on X, as X may be $\mu + \lambda\sigma$ and $\mu - \lambda\sigma$ with probability $1/2\lambda^2$ and otherwise μ. Note, however, that when X is a normal distribution with mean μ and standard deviation σ, then

$$\Pr[|X - \mu| \geq \lambda\sigma] = 2\int_{\lambda}^{\infty} \frac{1}{\sqrt{2\pi}} e^{-t^2/2}\, dt$$

and for λ large, this quantity is asymptotically $\sqrt{2/\pi}e^{-\lambda^2/2}/\lambda$, which is significantly smaller than $1/\lambda^2$. In Chapters 7 and 8 we shall see examples where X is the sum of "nearly independent" random variables and these better bounds can apply.

Suppose we have a decomposition

$$X = X_1 + \cdots + X_m.$$

Then var$[X]$ may be computed by the formula

$$\mathrm{var}[X] = \sum_{i=1}^{m} \mathrm{var}[X_i] + \sum_{i \neq j} \mathrm{cov}[X_i, X_j].$$

Here the second sum is over ordered pairs and the *covariance* cov$[Y,Z]$ is defined by

$$\mathrm{cov}[Y,Z] = E[YZ] - E[Y]E[Z].$$

In general, if Y,Z are independent, then cov$[Y,Z] = 0$. This often simplifies considerably variance calculations. Now suppose further, as will generally be the case in our applications, that the X_i are indicator random variables—that is, that $X_i = 1$ if a certain event A_i holds and otherwise $X_i = 0$. If X_i is one with probability $p_i = \Pr[A_i]$, then

$$\mathrm{var}[X_i] = p_i(1 - p_i) \leq p_i = E[X_i]$$

and so

$$\mathrm{var}[X] \leq E[X] + \sum_{i \neq j} \mathrm{cov}[X_i, X_j].$$

2. NUMBER THEORY

The second moment method is an effective tool in number theory. Let $\nu(n)$ denote the number of primes p dividing n. (We do not count multiplicity, though it would make little difference.) The following result says, roughly, that "almost all" n have "very close to" $\ln\ln n$ prime factors. This was first shown by Hardy and Ramanujan in 1920 by a quite complicated argument. We give the one-page proof of Paul Turán (1934), a proof that played a key role in the development of probabilistic methods in number theory.

Theorem 2.1. *Let $\omega(n) \to \infty$ arbitrarily slowly. Then the number of x in $\{1,\ldots,n\}$ such that*

$$|\nu(x) - \ln\ln n| > \omega(n)\sqrt{\ln\ln n}$$

is $o(n)$.

Proof. Let x be randomly chosen from $\{1,\ldots,n\}$. For p prime, set

$$X_p = \begin{cases} 1 & \text{if } p \mid x \\ 0 & \text{otherwise} \end{cases}$$

and set $X = \sum X_p$, the summation over all primes $p \leq n$, so that $X(x) = \nu(x)$. Now

$$E[X_p] = \frac{\lfloor n/p \rfloor}{n}.$$

As $y-1<\lfloor y\rfloor \leq y$,

$$E[X_p]=\frac{1}{p}+O\left(\frac{1}{n}\right).$$

By linearity of expectation (and using the prime number theorem),

$$E[X]=\sum_{p\leq n}\frac{1}{p}+O\left(\frac{1}{n}\right)=\ln\ln n+o(1).$$

Now we bound the variance

$$\operatorname{var}[X]\leq \ln\ln n+o(1)+\sum_{p\neq q}\operatorname{cov}[X_p,X_q].$$

With p,q distinct primes, $X_pX_q=1$ if and only if $p\mid x$ and $q\mid x$, which occurs if and only if $pq\mid x$. Hence

$$\begin{aligned}\operatorname{cov}[X_p,X_q]&=E[X_pX_q]-E[X_p]E[X_q]\\&=\frac{\lfloor n/pq\rfloor}{n}-\frac{\lfloor n/p\rfloor}{n}\frac{\lfloor n/q\rfloor}{n}\\&\leq\frac{1}{pq}-\left(\frac{1}{p}-\frac{1}{n}\right)\left(\frac{1}{q}-\frac{1}{n}\right)\leq\frac{1}{n}\left(\frac{1}{p}+\frac{1}{q}\right).\end{aligned}$$

Thus

$$\sum_{p\neq q}\operatorname{cov}[X_p,X_q]\leq\frac{1}{n}\sum_{p\neq q}\left(\frac{1}{p}+\frac{1}{q}\right)=\frac{\pi(n)-1}{n}\sum_{p}\frac{2}{p},$$

where $\pi(n)\sim n/\ln n$ is the number of primes $p\leq n$. So

$$\sum_{p\neq q}\operatorname{cov}[X_p,X_q]<\frac{(n/\ln n)}{n}(2\ln\ln n)=o(1).$$

That is, the covariances do not affect the variance, $\operatorname{var}[X]=\ln\ln n+o(1)$, and Chebyschev's inequality actually gives

$$\Pr\left[|\nu(n)-\ln\ln n|>\lambda\sqrt{\ln\ln n}\right]<\lambda^{-2}+o(1)$$

for any constant λ. ∎

In a classic paper, Paul Erdős and Marc Kac (1940) showed, essentially, that X does behave like a normal distribution with mean and variance $\ln\ln n+o(1)$. Here is their precise result.

The Erdős–Kac Theorem. *Let λ be fixed, positive, negative, or zero. Then*

$$\lim_{n\to\infty}\frac{1}{n}\left|\left\{x:1\leq x\leq n,\nu(x)\geq\ln\ln n+\lambda\sqrt{\ln\ln n}\right\}\right|=\int_{\lambda}^{\infty}\frac{1}{\sqrt{2\pi}}e^{-t^2/2}\,dt.$$

We do not prove this result here.

3. MORE BASICS

Let X be a nonnegative integral valued random variable and suppose we want to bound $\Pr[X = 0]$ given the value $\mu = E[X]$. If $\mu < 1$, we may use the inequality

$$\Pr[X > 0] \leq E[X]$$

so that if $E[X] \to 0$ then $X = 0$ almost always. (Here we are imagining an infinite sequence of X dependent on some parameter n going to infinity.) But now suppose $E[X] \to \infty$. It does *not* necessarily follow that $X > 0$ almost always. For example, let X be the number of deaths due to nuclear war in the 12 months after reading this paragraph. Calculation of $E[X]$ can make for lively debate but few would deny that it is quite large. Yet we may believe—or hope—that $\Pr[X \neq 0]$ is very close to 0. We can sometimes deduce $X > 0$ almost always if we have further information about $\mathrm{var}[X]$.

Theorem 3.1.

$$Pr[X = 0] \leq \frac{\mathrm{var}[X]}{E[X]^2}.$$

Proof. Set $\lambda = \mu/\sigma$ in Chebyschev's inequality. Then

$$\Pr[X = 0] \leq \Pr[|X - \mu| \geq \lambda\sigma] \leq \frac{1}{\lambda^2} = \frac{\sigma^2}{\mu^2}. \qquad \blacksquare$$

We generally apply this result in asymptotic terms.

Corollary 3.2. *If* $E[X] \to \infty$ *and* $\mathrm{var}[X] = o(E[X]^2)$, *then* $X > 0$ *a.a.*

The proof of Theorem 3.1 actually gives that for any $\epsilon > 0$,

$$\Pr[|X - E[X]| \geq \epsilon E[X]] \leq \frac{\mathrm{var}[X]}{\epsilon^2 E[X]^2}$$

and thus in asymptotic terms we actually have the following stronger assertion.

Corollary 3.3. *If* $\mathrm{var}[X] = o(E[X]^2)$, *then* $X \sim E[X]$ *a.a.*

Suppose again $X = X_1 + \cdots + X_m$ where X_i is the indicator random variable for event A_i. For indices i, j, write $i \sim j$ if $i \neq j$ and the events A_i, A_j are not independent. We set (the sum over ordered pairs)

$$\Delta = \sum_{i \sim j} \Pr[A_i \wedge A_j].$$

Note that when $i \sim j$,

$$\mathrm{cov}[X_i, X_j] = E[X_i X_j] - E[X_i]E[X_j] \leq E[X_i X_j] = \Pr[A_i \wedge A_j]$$

and that when $i \neq j$ and not $i \sim j$, then $\text{cov}[X_i, X_j] = 0$. Thus

$$\text{var}[X] \leq E[X] + \Delta.$$

Corollary 3.4. *If $E[X] \to \infty$ and $\Delta = o(E[X]^2)$, then $X > 0$ almost always. Furthermore $X \sim E[X]$ almost always.*

Let us say $X_1, \ldots, X_m$ are *symmetric* if for every $i \neq j$ there is an automorphism of the underlying probability space that sends event A_i to event A_j. Examples will appear in the next section. In this instance, we write

$$\Delta = \sum_{i \sim j} \Pr[A_i \wedge A_j] = \sum_i \Pr[A_i] \sum_{j \sim i} \Pr[A_j \mid A_i]$$

and note that the inner summation is independent of i. We set

$$\Delta^* = \sum_{j \sim i} \Pr[A_j \mid A_i],$$

where i is any fixed index. Then

$$\Delta = \sum_i \Pr[A_i] \Delta^* = \Delta^* \sum_i \Pr[A_i] = \Delta^* E[X].$$

Corollary 3.5. *If $E[X] \to \infty$ and $\Delta^* = o(E[X])$, then $X > 0$ almost always. Furthermore $X \sim E[X]$ almost always.*

The condition of Corollary 3.5 has the intuitive sense that conditioning on any specific A_i holding does not substantially increase the expected number $E[X]$ of events holding.

4. RANDOM GRAPHS

The definition of the random graph $G(n,p)$ and of "threshold function" are given in Chapter 10, Section 1. The results of this section are generally surpassed by those of Chapter 10, but they were historically the first results and provide a good illustration of the second moment. We begin with a particular example. By $\omega(G)$, we denote here and in the rest of the book the number of vertices in the maximum clique of the graph G.

Theorem 4.1. *The property $\omega(G) \geq 4$ has threshold function $n^{-2/3}$.*

Proof. For every 4-set S of vertices in $G(n,p)$ let A_S be the event "S is a clique" and X_S its indicator random variable. Then

$$E[X_S] = \Pr[A_S] = p^6,$$

as six different edges must all lie in $G(n,p)$. Set

$$X = \sum_{|S|=4} X_S$$

so that X is the number of 4-cliques in G and $\omega(G) \geq 4$ if and only if $X > 0$. Linearity of expectation gives

$$E[X] = \sum_{|S|=4} E[X_S] = \binom{n}{4} p^6 \sim \frac{n^4 p^6}{24}.$$

When $p(n) \ll n^{-2/3}$, $E[X] = o(1)$, and so $X = 0$ almost surely.

Now suppose $p(n) \gg n^{-2/3}$ so that $E[X] \to \infty$ and consider the Δ^* of Corollary 3.5. (All 4-sets "look the same" so that the X_S are symmetric.) Here $S \sim T$ if and only if $S \neq T$ and S,T have common edges—that is, if and only if $|S \cap T| = 2$ or 3. Fix S. There are $O(n^2)$ sets T with $|S \cap T| = 2$ and for each of these $\Pr[A_T \mid A_S] = p^5$. There are $O(n)$ sets T with $|S \cap T| = 3$ and for each of these $\Pr[A_T \mid A_S] = p^3$. Thus

$$\Delta^* = O(n^2 p^5) + O(np^3) = o(n^4 p^6) = o(E[X]),$$

since $p \gg n^{-2/3}$. Corollary 3.5 therefore applies and $X > 0$, i.e., there *does* exist a clique of size 4, almost always. ■

The proof of Theorem 4.1 appears to require a fortuitous calculation of Δ^*. The following definitions will allow for a description of when these calculations work out.

Definitions. Let H be a graph with v vertices and e edges. We call $\rho(H) = e/v$ the *density* of H. We call H *balanced* if every subgraph H' has $\rho(H') \leq \rho(H)$. We call H *strictly balanced* if every proper subgraph H' has $\rho(H') < \rho(H)$.

Examples. K_4 and, in general, K_k are strictly balanced. The graph

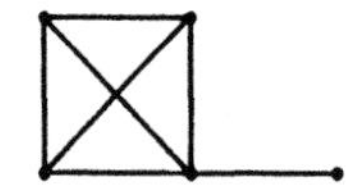

is not balanced, as it has density 7/5, while the subgraph K_4 has density 3/2. The graph

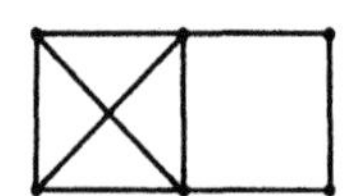

is balanced, but not strictly balanced as it and its subgraph K_4 have density 3/2. ■

Theorem 4.2. *Let H be a balanced graph with v vertices and e edges. Let $A(G)$ be the event that H is a subgraph (not necessarily induced) of G. Then $p = n^{-v/e}$ is the threshold function for A.*

Proof. We follow the argument of Theorem 4.1. For each v-set S, let A_S be the event that $G|_S$ contains H as a subgraph. Then

$$p^e \le \Pr[A_S] \le v!\,p^e.$$

(Any particular placement of H has probability p^e of occurring and there are at most $v!$ possible placements. The precise calculation of $\Pr[A_S]$ is, in general, complicated due to the overlapping of potential copies of H.) Let X_S be the indicator random variable for A_S and

$$X = \sum_{|S|=v} X_S$$

so that A holds if and only if $X > 0$. Linearity of expectation gives

$$E[X] = \sum_{|S|=v} E[X_S] = \binom{n}{v}\Pr[A_S] = \Theta(n^v p^e).$$

If $p \ll n^{-v/e}$, then $E[X] = o(1)$, so $X = 0$ almost always.

Now assume $p \gg n^{-v/e}$ so that $E[X] \to \infty$ and consider the Δ^* of Corollary 3.5. (All v-sets look the same so the X_S are symmetric.) Here $S \sim T$ if and only if $S \neq T$ and S,T have common edges—that is, if and only if $|S \cap T| = i$ with $2 \le i \le v-1$. Let S be fixed. We split

$$\Delta^* = \sum_{T \sim S} \Pr[A_T \mid A_S]$$

$$= \sum_{i=2}^{v-1} \sum_{|T \cap S|=i} \Pr[A_T \mid A_S].$$

For each i, there are $O(n^{v-i})$ choices of T. Fix S,T and consider $\Pr[A_T \mid A_S]$. There are $O(1)$ possible copies of H on T. Each has—since, critically, H is balanced—at most ie/v edges with both vertices in S and thus at least $e - ie/v$ other edges. Hence

$$\Pr[A_T \mid A_S] = O(p^{e-ie/v})$$

and

$$\Delta^* = \sum_{i=2}^{v-1} O(n^{v-i} p^{e-ie/v})$$

$$= \sum_{i=2}^{v-1} O((n^v p^e)^{1-i/v})$$

$$= \sum_{i=2}^{v-1} o(n^v p^e)$$

$$= o(E[X]),$$

since $n^v p^e \to \infty$. Therefore Corollary 3.5 applies. ∎

Theorem 4.3. *In the notation of Theorem 4.2, if H is not balanced, then $p = n^{-v/e}$ is not the threshold function for A.*

Proof. Let H_1 be a subgraph of H with v_1 vertices, e_1 edges, and $e_1/v_1 > e/v$. Let α satisfy $v_1/e_1 < \alpha < v/e$ and set $p = n^{-\alpha}$. The expected number of copies of H_1 is then $o(1)$, so almost always $G(n,p)$ contains no copy of H_1. But if it contains no copy of H_1, then it surely can contain no copy of H. ∎

The threshold function for the property of containing a copy of H, for general H, can be determined in a similar way. Let H_1 be that subgraph with maximal density $\rho(H_1) = e_1/v_1$. (When H is balanced we may take $H_1 = H$.) It can be shown that $p = n^{-v_1/e_1}$ is the threshold function. We do not show this here though it follows fairly straightforwardly from Theorem 4.5 below.

We finish this section with two strengthenings of Theorem 4.2.

Theorem 4.4. *Let H be strictly balanced with v vertices, e edges, and a automorphisms. Let X be the number of copies of H in $G(n,p)$. Assume $p \gg n^{-v/e}$. Then almost always*

$$X \sim \frac{n^v p^e}{a}.$$

Proof. Label the vertices of H by $1,\ldots,v$. For each ordered $x_1,\ldots,x_v$, let $A_{x_1,\ldots,x_v}$ be the event that $x_1,\ldots,x_v$ provides a copy of H in that order. Specifically, we define

$$A_{x_1,\ldots,x_v} : \{i,j\} \in E(H) \to \{x_i, x_j\} \in E(G).$$

We let $I_{x_1,\ldots,x_v}$ be the corresponding indicator random variable. We define an equivalence class on v-tuples by setting $(x_1,\ldots,x_v) \equiv (y_1,\ldots,y_v)$ if there is an automorphism σ of $V(H)$ so that $y_{\sigma(i)} = x_i$ for $1 \le i \le v$. Then

$$X = \sum I_{x_1,\ldots,x_v}$$

gives the number of copies of H in G where the sum is taken over one entry from each equivalence class. As there are $(n)_v/a$ terms,

$$E[X] = \frac{(n)_v}{a} E[I_{x_1,\ldots,x_v}] = \frac{(n)_v p^e}{a} \sim \frac{n^v p^e}{a}.$$

Our assumption $p \gg n^{-v/e}$ implies $E[X] \to \infty$. It suffices therefore to show $\Delta^* = o(E[X])$. Fixing $x_1,\ldots,x_v$,

$$\Delta^* = \sum_{(y_1,\ldots,y_v)\sim(x_1,\ldots,x_v)} \Pr\left[A_{(y_1,\ldots,y_v)} \mid A_{(x_1,\ldots,x_v)}\right].$$

There are $v!/a = O(1)$ terms with $\{y_1,\ldots,y_v\} = \{x_1,\ldots,x_v\}$ and for each the conditional probability is at most 1 (actually, at most p), thus contributing $O(1) = o(E[X])$ to Δ^*. When $\{y_1,\ldots,y_v\} \cap \{x_1,\ldots,x_v\}$ has i elements, $2 \leq i \leq v-1$, the argument of Theorem 4.2 gives that the contribution to Δ^* is $o(E[X])$. Altogether, $\Delta^* = o(E[X])$ and we apply Corollary 3.5. ■

Theorem 4.5. *Let H be any fixed graph. For every subgraph H' of H (including H itself), let $X_{H'}$ denote the number of copies of H' in $G(n,p)$. Assume p is such that $E[X_{H'}] \to \infty$ for every H'. Then*

$$X_H \sim E[X_H]$$

almost always.

Proof. Let H have v vertices and e edges. As in Theorem 4.4, it suffices to show $\Delta^* = o(E[X])$. We split Δ^* into a finite number of terms. For each H' with w vertices and f edges, we have those $(y_1,\ldots,y_v)$ that overlap with the fixed $(x_1,\ldots,x_v)$ in a copy of H'. These terms contribute, up to constants,

$$n^{v-w} p^{e-f} = \Theta\left(\frac{E[X_H]}{E[X_{H'}]}\right) = o(E[X_H])$$

to Δ^*. Hence Corollary 3.5 does apply. ■

5. CLIQUE NUMBER

Now we fix edge probability $p = 1/2$ and consider the clique number $\omega(G)$. We set

$$f(k) = \binom{n}{k} 2^{-\binom{k}{2}},$$

the expected number of k-cliques. The function $f(k)$ drops under one at $k \sim 2\log_2 n$. (Very roughly, $f(k)$ is like $n^k 2^{-k^2/2}$.)

Theorem 5.1. *Let $k = k(n)$ satisfy $k \sim 2\log_2 n$ and $f(k) \to \infty$. Then almost always $\omega(G) \geq k$.*

Proof. For each k-set S, let A_S be the event "S is a clique" and X_S the corresponding indicator random variable. We set

$$X = \sum_{|S|=k} X_S$$

so that $\omega(G) \geq k$ if and only if $X > 0$. Then $E[X] = f(k) \to \infty$ and we examine the Δ^* of Corollary 3.5. Fix S and note that $T \sim S$ if and only if $|T \cap S| = i$ where $2 \leq i \leq k-1$. Hence

$$\Delta^* = \sum_{i=2}^{k-1} \binom{k}{i} \binom{n-k}{k-i} 2^{\binom{i}{2}-\binom{k}{2}}$$

and so

$$\frac{\Delta^*}{E[X]} = \sum_{i=2}^{k-1} g(i),$$

where we set

$$g(i) = \frac{\binom{k}{i}\binom{n-k}{k-i}}{\binom{n}{k}} 2^{\binom{i}{2}}.$$

Observe that $g(i)$ may be thought of as the probability that a randomly chosen T will intersect a fixed S in i points times the factor increase in $\Pr[A_T]$ when it does. Setting $i = 2$,

$$g(2) = 2\frac{\binom{k}{2}\binom{n-k}{k-2}}{\binom{n}{k}} \sim \frac{k^4}{n^2} = o(1).$$

At the other extreme, $i = k-1$,

$$g(k-1) = \frac{k(n-k)2^{-(k-1)}}{\binom{n}{k}2^{-\binom{k}{2}}} \sim \frac{2kn2^{-k}}{E[X]}.$$

As $k \sim 2\log_2 n$, the numerator is $n^{-1+o(1)}$. The denominator approaches infinity and so $g(k-1) = o(1)$. Some detailed calculation (which we omit) gives that the remaining $g(i)$ are also negligible so that Corollary 3.5 applies. ∎

Theorem 5.1 leads to a strong concentration result for $\omega(G)$. For $k \sim 2\log_2 n$,

$$\frac{f(k+1)}{f(k)} = \frac{n-k+1}{k+1} 2^{-k} = n^{-1+o(1)} = o(1).$$

Let $k_0 = k_0(n)$ be that value with $f(k_0) \geq 1 > f(k_0+1)$. For "most" n, the function $f(k)$ will jump from a large $f(k_0)$ to a small $f(k_0+1)$. The probability that G contains a clique of size k_0+1 is at most $f(k_0+1)$, which will be quite small. When $f(k_0)$ is large, Theorem 5.1 implies that G contains a clique of size k_0 with probability nearly 1. Together, with very high probability, $\omega(G) = k_0$. For some n, one of the values $f(k_0), f(k_0+1)$ may be of moderate size so this argument does not apply. Still one may show a strong concentration result found independently by Bollobás and Erdős (1976) and Matula (1976). (See Chapter 10 for more details.)

Corollary 5.2. *There exists $k = k(n)$ so that*

$$\Pr[\omega(G) = k \text{ or } k+1] \to 1.$$

6. DISTINCT SUMS

A set $x_1, \ldots, x_k$ of positive integers is said to have distinct sums if all sums

$$\sum_{i \in S} x_i, \qquad S \subseteq \{1, \ldots, k\},$$

are distinct. Let $f(n)$ denote the maximal k for which there exists a set

$$\{x_1, \ldots, x_k\} \subset \{1, \ldots, n\}$$

with distinct sums. The simplest example of a set with distinct sums is $\{2^i : i \le \log_2 n\}$. This example shows

$$f(n) \ge 1 + \lfloor \log_2 n \rfloor.$$

Erdős has offered \$300 for a proof or disproof that

$$f(n) \le \log_2 n + C$$

for some constant C. From above, as all $2^{f(n)}$ sums are distinct and less than nk,

$$2^{f(n)} < nk$$

and so

$$f(n) < \log_2 n + \log_2 \log_2 n + O(1).$$

Examination of the second moment gives a modest improvement. Fix $\{x_1, \ldots, x_k\} \subset \{1, \ldots, n\}$ with distinct sums. Let $\epsilon_1, \ldots, \epsilon_k$ be independent with

$$\Pr[\epsilon_i = 1] = \Pr[\epsilon_i = 0] = \tfrac{1}{2}$$

and set

$$X = \epsilon_1 x_1 + \cdots + \epsilon_k x_k.$$

(We may think of X as a random sum.) Set

$$\mu = E[X] = \frac{x_1 + \cdots + x_k}{2}$$

and $\sigma^2 = \mathrm{var}[X]$. We bound

$$\sigma^2 = \frac{x_1^2 + \cdots + x_k^2}{4} \le \frac{n^2 k}{4}$$

so that $\sigma \leq n\sqrt{k}/2$. By Chebyschev's inequality, for any $\lambda > 1$,

$$\Pr\left[|X - \mu| \geq \frac{\lambda n\sqrt{k}}{2}\right] \leq \lambda^{-2}.$$

Reversing,

$$1 - \frac{1}{\lambda^2} \leq \Pr\left[|X - \mu| < \frac{\lambda n\sqrt{k}}{2}\right].$$

But X has any particular value with probability either 0 or 2^{-k}, since, critically, a sum can be achieved in at most one way. Thus

$$\Pr\left[|X - \mu| < \frac{\lambda n\sqrt{k}}{2}\right] \leq 2^{-k}\lambda n\sqrt{k}$$

and

$$n \geq \frac{2^k}{\sqrt{k}} \frac{1 - \lambda^{-2}}{\lambda}.$$

While $\lambda = \sqrt{3}$ gives optimal results, any choice of $\lambda > 1$ gives the following.

Theorem 6.1.

$$f(n) \leq \log_2 n + \tfrac{1}{2}\log_2\log_2 n + O(1).$$

The Probabilistic Lens: Hamiltonian Paths

What is the maximum possible number of directed Hamiltonian paths in a tournament on n vertices? Denote this number by $P(n)$. The first application of the probabilistic method in combinatorics is the result of Szele (1943) described in Chapter 2, which states that $P(n) \geq n!/2^{n-1}$. This bound follows immediately from the observation that the right-hand side is the expected number of such paths in a random tournament on n vertices. In the same paper, Szele shows that

$$\frac{1}{2} \leq \lim_{n\to\infty} \left(\frac{P(n)}{n!}\right)^{1/n} \leq \frac{1}{2^{3/4}},$$

proves that this limit does exist, and conjectures that its correct value is 1/2.

This conjecture is proved in Alon (1990b). The proof is given below. The main tool is Brégman's proof of Minc conjecture for the permanent of a $(0,1)$-matrix, described in "The Probabilistic Lens: Brégman's Theorem" (following Chapter 2).

Theorem. *There exists a positive constant c such that for every n,*

$$P(n) \leq cn^{3/2}\frac{n!}{2^{n-1}}.$$

Proof. For a tournament T, denote by $P(T)$ the number of directed Hamiltonian paths of T. Similarly, $C(T)$ denotes the number of directed Hamiltonian cycles of T, and $F(T)$ denotes the number of spanning subgraphs of T in which the indegree and the outdegree of every vertex is exactly 1. Clearly,

$$C(T) \leq F(T). \tag{1}$$

If $T = (V,E)$ is a tournament on a set $V = \{1,2,\ldots,n\}$ of n vertices, the *adjacency matrix* of T is the n by n $(0,1)$-matrix $A_T = (a_{ij})$ defined by $a_{ij} = 1$ if $(i,j) \in E$ and $a_{ij} = 0$ otherwise. Let r_i denote the number of ones in row i. Clearly,

$$\sum_i^n r_i = \binom{n}{2}. \tag{2}$$

By interpreting combinatorially the terms in the expansion of the permanent $\text{per}(A_T)$, it follows that

$$\text{per}(A_T) = F(T). \tag{3}$$

We need the following technical lemma.

Lemma. *For every two integers a, b satisfying $b \geq a + 2 > a \geq 1$, the inequality*

$$(a!)^{1/a} \cdot (b!)^{1/b} < ((a+1)!)^{1/(a+1)} \cdot ((b-1)!)^{1/(b-1)}$$

holds.

Proof. The assertion is simply that $f(a) < f(b-1)$, for the function f defined by

$$f(a) = \frac{(a!)^{1/a}}{((a+1)!)^{1/(a+1)}}.$$

Thus it suffices to show that for every integer $x \geq 2$, $f(x-1) < f(x)$. Substituting the expression for f and raising both sides to the power $x(x-1)(x+1)$, it follows that it suffices to show that for all $x \geq 2$,

$$((x-1)!)^{x(x+1)} \cdot ((x+1)!)^{x(x-1)} < (x!)^{2(x^2-1)},$$

that is,

$$\left(\frac{x^x}{x!}\right)^2 > \left(\frac{x+1}{x}\right)^{x(x-1)}.$$

This is certainly true for $x = 2$. For $x \geq 3$, it follows from the facts that $4^x > e^{x+1}$, that $x! < ((x+1)/2)^x$ and that $e^{x-1} > ((x+1)/x)^{x(x-1)}$. ■

Corollary. *Define $g(x) = (x!)^{1/x}$. For every integer $S \geq n$, the maximum of the function $\prod_{i=1}^{n} g(x_i)$ subject to the constraints $\sum_{i=1}^{n} x_i = S$ and $x_i \geq 1$ are integers, is obtained iff the variables x_i are as equal as possible (i.e., iff each x_i is either $\lfloor S/n \rfloor$ or $\lceil S/n \rceil$.)*

Proof. If there are two indices i and j such that $x_i \geq x_j + 2$, then, by the above lemma, the value of the product would increase once we add 1 to x_j and subtract 1 from x_i. ■

Returning to our tournament T, we observe that the numbers r_i defined above are precisely the outdegrees of the vertices of T. If at least one of these is 0, then clearly $C(T) = F(T) = 0$. Otherwise, by Brégman's theorem, by the above corollary, and by (2) and (3), $F(T)$ is at most the value of the function $\prod_{i=1}^{n}(r_i!)^{1/r_i}$, where the integral variables r_i satisfy (2) and are as equal as possible. By a straightforward (though somewhat tedious) derivation of the asymptotics using Stirling's formula, this gives:

Proposition. *For every tournament T on n vertices,*

$$C(T) \leq F(T) \leq (1+o(1))\frac{\sqrt{\pi}}{\sqrt{2}e}n^{3/2}\frac{(n-1)!}{2^n}.$$

To complete the proof of the theorem, we have to derive a bound for the number of Hamiltonian paths in a tournament from the above result. Given a tournament S on n vertices, let T be the random tournament ontained from S by adding to it a new vertex y and by orienting each edge connecting y with one of the vertices of S, randomly and independently. For every fixed Hamiltonian path in S, the probability that it can be extended to a Hamiltonian cycle in T is precisely 1/4. Thus the expected number of Hamiltonian cycles in T is $1/4\ P(S)$ and hence there is a specific T for which $C(T) \geq 1/4\ P(S)$. However, by the above proposition,

$$C(T) \leq (1+o(1))\frac{\sqrt{\pi}}{\sqrt{2}e}(n+1)^{3/2}\frac{(n+1)!}{2^n},$$

and therefore

$$P(T) \leq O\left(n^{3/2}\frac{n!}{2^{n-1}}\right),$$

completing the proof of the theorem. ■

5

The Local Lemma

1. THE LEMMA

In a typical probabilistic proof of a combinatorial result, one usually has to show that the probability of a certain event is positive. However, many of these proofs actually give more and show that the probability of the event considered is not only positive but is large. In fact, most probabilistic proofs deal with events that hold with high probability, that is, a probability that tends to 1 as the dimensions of the problem grow. For example, consider the proof given in Chapter 1 that for each $k \geq 1$ there are tournaments in which for every set of k players there is one who beats them all. The proof actually shows that for every fixed k, if the number n of players is sufficiently large, then almost all tournaments with n players satisfy this property, that is, the probability that a random tournament with n players has the desired property tends to 1 as n tends to infinity.

On the other hand, there is a trivial case in which one can show that a certain event holds with positive, though very small, probability. Indeed, if we have n mutually independent events and each of them holds with probability at least $p > 0$, then the probability that all events hold simultaneously is at least p^n, which is positive, although it may be exponentially small in n.

It is natural to expect that the case of mutual independence can be generalized to that of rare dependencies, and provide a more general way of proving that certain events hold with positive, though small, proability. Such a generalization is, indeed, possible, and is stated in the following lemma, known as the Lovász local lemma. This simple lemma, first proved in (Erdős and Lovász [1975]) is an extremely powerful tool, as it supplies a way for dealing with rare events.

Lemma 1.1 (The Local Lemma; General Case). *Let $A_1, A_2, \ldots, A_n$ be events in an arbitrary probability space. A directed graph $D = (V, E)$ on the set of vertices $V = \{1, 2, \ldots, n\}$ is called a dependency digraph for the events $A_1, \ldots,$ A_n if for each i, $1 \leq i \leq n$, the event A_i is mutually independent of all the events $\{A_j : (i, j) \notin E\}$. Suppose that $D = (V, E)$ is a dependency digraph for the above events and suppose there are real numbers $x_1, \ldots, x_n$ such that $0 \leq$*

$x_i < 1$ and

$$\Pr(A_i) \leq x_i \prod_{(i,j)\in E} (1 - x_j)$$

for all $1 \leq i \leq n$. Then

$$\Pr\left(\bigwedge_{i=1}^{n} \overline{A}_i\right) \geq \prod_{i=1}^{n}(1 - x_i).$$

In particular, with positive probability no event A_i holds.

Proof. We first prove, by induction on s, that for any $S \subset \{1,\ldots,n\}$, $|S| = s < n$, and $i \notin S$,

$$\Pr\left(A_i \middle| \bigwedge_{j\in S} \overline{A}_j\right) \leq x_i. \tag{1}$$

This is certainly true for $s = 0$. Assuming it holds for all $s' < s$, we prove it for S. Put $S_1 = \{j \in S; (i,j) \in E\}$, $S_2 = S \backslash S_1$. Then

$$\Pr\left(A_i \middle| \bigwedge_{j\in S} \overline{A}_j\right) = \frac{\Pr\left(A_i \wedge \left(\bigwedge_{j\in S_1} \overline{A}_j\right) \mid \bigwedge_{\ell\in S_2} \overline{A}_\ell\right)}{\Pr\left(\bigwedge_{j\in S_1} \overline{A}_j \mid \bigwedge_{\ell\in S_2} \overline{A}_\ell\right)}. \tag{2}$$

To bound the numerator, observe that since A_i is mutually independent of the events $\{A_\ell : \ell \in S_2\}$,

$$\Pr\left(A_i \wedge \left(\bigwedge_{j\in S_1} \overline{A}_j\right) \middle| \bigwedge_{\ell\in S_2} \overline{A}_\ell\right) \leq \Pr\left(A_i \middle| \bigwedge_{\ell\in S_2} \overline{A}_\ell\right) = \Pr(A_i) \leq x_i \prod_{(i,j)\in E} (1 - x_j). \tag{3}$$

The denominator, on the other hand, can be bounded by the induction hypothesis. Indeed, suppose $S_1 = \{j_1, j_2, \ldots, j_r\}$. If $r = 0$, then the denominator is 1, and (1) follows. Otherwise

$$\begin{aligned}
\Pr\left(\overline{A}_{j_1} \wedge \overline{A}_{j_2} \wedge \cdots \wedge \overline{A}_{j_r} \middle| \bigwedge_{\ell\in S_2} \overline{A}_\ell\right) &= \left(1 - \Pr(A_{j_1} \middle| \bigwedge_{\ell\in S_2} \overline{A}_\ell)\right) \\
&\quad \cdot \left(1 - \Pr\left(A_{j_2} \middle| \overline{A}_{j_1} \wedge \bigwedge_{\ell\in S_2} \overline{A}_\ell\right)\right) \cdots \\
&\quad \cdot \left(1 - \Pr\left(A_{j_r} \mid \overline{A}_{j_1} \wedge \cdots \wedge \overline{A}_{j_{r-1}} \wedge \bigwedge_{\ell\in S_2} \overline{A}_\ell\right)\right) \\
&\geq (1 - x_{j_1})(1 - x_{j_2}) \cdots (1 - x_{j_r}) \geq \prod_{(i,j)\in E} (1 - x_j).
\end{aligned} \tag{4}$$

Substituting (3) and (4) into (2), we conclude that $\Pr(A_i \mid \bigwedge_{j\in S}\overline{A}_j) \le x_i$, completing the proof of the induction.

The assertion of Lemma 1.1 now follows easily, as

$$\Pr\left(\bigwedge_{i=1}^{n}\overline{A}_i\right) = (1-\Pr(A_1))\cdot(1-\Pr(A_2\mid\overline{A}_1))\cdot\cdots$$

$$\cdot\left(1-\Pr\left(A_n\,\middle|\,\bigwedge_{i=1}^{n-1}\overline{A}_i\right)\right) \ge \prod_{i=1}^{n}(1-x_i),$$

completing the proof. ∎

Corollary 1.2 (The Local Lemma; Symmetric Case). *Let $A_1, A_2, \ldots, A_n$ be events in an arbitrary probability space. Suppose that each event A_i is mutually independent of a set of all the other events A_j but at most d, and that* $\Pr(A_i) \le p$ *for all* $1 \le i \le n$*. If*

$$ep(d+1) \le 1, \tag{5}$$

then $\Pr(\bigwedge_{i=1}^{n}\overline{A}_i) > 0$.

Proof. If $d = 0$, the result is trivial. Otherwise, by the assumption, there is a dependency digraph $D = (V, E)$ for the events $A_1, \ldots, A_n$ in which for each $i, |\{j : (i,j) \in E\}| \le d$. The result now follows from Lemma 1.1 by taking $x_i = 1/(d+1)(< 1)$ for all i and using the fact that for any $d \ge 2$, $[1-1/(d+1)]^d > 1/e$. ∎

It is worth noting that as shown by Shearer (1985), the constant "e" is the best possible constant in inequality (5). Note also that the proof of Lemma 1.1 indicates that the conclusion remains true even when we replace the two assumptions that each A_i is mutually independent of $\{A_j : (i,j) \notin E)$ and that

$$\Pr(A_i) \le x_i \prod_{(ij)\in E}(1-x_j)$$

by the weaker assumption that for each i and each $S_2 \subset \{1, \ldots, n\}\backslash\{j : (i,j) \in E\}$,

$$\Pr\left(x_i\,\middle|\,\bigwedge_{j\in S_2}\overline{A}_j\right) \le x_i \prod_{(i,j)\in E}(1-x_j).$$

This turns out to be useful in certain applications.

In the next few sections, we present various applications of the local lemma for obtaining combinatorial results. There is no known proof of any of these results, which does not use the local lemma.

2. PROPERTY B AND MULTICOLORED SETS OF REAL NUMBERS

Recall that a hypergraph $H = (V,E)$ has property B, (i.e., is 2-colorable), if there is a coloring of V by two colors so that no edge $f \in E$ is monochromatic.

Theorem 2.1. *Let $H = (V,E)$ be a hypergraph in which every edge has at least k elements, and suppose that each edge of H intersects at most d other edges. If $e(d+1) \leq 2^{k-1}$, then H has property B.*

Proof. Color each vertex v of H, randomly and independently, either blue or red (with equal probability). For each edge $f \in E$, let A_f be the event that f is monochromatic. Clearly, $\Pr(A_f) = 2/2^{|f|} \leq 1/2^{k-1}$. Moreover, each event A_f is clearly mutually independent of all the other events $A_{f'}$ for all edges f' that do not intersect f. The result now follows from Corollary 1.2. ■

A special case of Theorem 2.1 is that for any $k \geq 9$, any k-uniform k-regular hypergraph H has property B. Indeed, since any edge f of such an H contains k vertices, each of which is incident with k edges (including f), it follows that f intersects at most $d = k(k-1)$ other edges. The desired result follows, since $e(k(k-1)+1) < 2^{k-1}$ for each $k \geq 9$. This special case has a different proof (see Alon and Bregman [1988]), which works for each $k \geq 8$. It is plausible to conjecture that in fact for each $k \geq 4$, each k-uniform k-regular hypergraph is 2-colorable.

The next result we consider, which appeared in the original paper of Erdős and Lovász (1975), deals with k-colorings of the real numbers. For a k-coloring $c : \mathbf{R} \to \{1,2,\ldots,k\}$ of the real numbers by the k colors $1,2,\ldots,k$, and for a subset $T \subset \mathbf{R}$, we say that T is *multicolored* (with respect to c) if $c(T) = \{1,2,\ldots,k\}$, that is, if T contains elements of all colors.

Theorem 2.2. *Let m and k be two positive integers satisfying*

$$e\,(m(m-1)+1)k\left(1-\frac{1}{k}\right)^m \leq 1. \tag{6}$$

Then, for any set S of m real numbers, there is a k-coloring so that each translation $x + S$ (for $x \in \mathbf{R}$) is multicolored.

Notice that (6) holds whenever $m > (3+o(1))k \log k$.

Proof. We first fix a *finite* subset $X \subseteq \mathbf{R}$ and show the existence of a k-coloring so that each translation $x + S$ (for $x \in X$) is multicolored. This is an easy consequence of the local lemma. Indeed, put $Y = \bigcup_{x \in X}(x + S)$ and let $c : Y \to \{1,2,\ldots,k\}$ be a random k-coloring of Y obtained by choosing, for each $y \in Y$, randomly and independently, $c(y) \in \{1,2,\ldots,k\}$ according to a uniform distribution on $\{1,2,\ldots,k\}$. For each $x \in X$, let A_x be the event that $x+S$ is not multicolored (with respect to c). Clearly, $\Pr(A_x) \leq k(1-1/k)^m$. Moreover, each event A_x is mutually independent of all the other events $A_{x'}$ but those for which $(x+S)\cap(x'+S) \neq \emptyset$. As there are at most $m(m-1)$ such events, the desired result follows from Corollary 1.2.

We can now prove the existence of a coloring of the set of all reals with the desired properties, by a standard compactness argument. Since the discrete space with k points is (trivially) compact, Tihonov's theorem (which is equivalent to the axiom of choice) implies that an arbitrary product of such spaces is compact. In particular, the space of all functions from $\mathbf{R}$ to $\{1,2,\ldots,k\}$, with the usual product topology, is compact. In this space for every fixed $x \in \mathbf{R}$, the set C_x of all colorings c, such that $x+S$ is multicolored is closed. (In fact, it is both open and closed, since a basis to the open sets is the set of all colorings whose values are prescribed in a finite number of places.) As we proved above, the intersection of any finite number of sets C_x is nonempty. It thus follows, by compactness, that the intersection of all sets C_x is nonempty. Any coloring in this intersection has the properties in the conclusion of Theorem 2.2. ■

Note that it is impossible, in general, to apply the local lemma to an infinite number of events and conclude that in some point of the probability space none of them holds. In fact, there are trivial examples of countably many mutually independent events A_i, satisfying $\Pr(A_i) = 1/2$ and $\bigwedge_{i\geq 1} \overline{A_i} = \emptyset$. Thus the compactness argument is essential in the above proof.

3. LOWER BOUNDS FOR RAMSEY NUMBERS

The derivation of lower bounds for Ramsey numbers by Erdős in 1947 was one of the first applications of the probabilistic method. The local lemma provides a simple way of improving these bounds. Let us obtain, first, a lower bound for the diagonal Ramsey number $R(k,k)$. Consider a random 2-coloring of the edges of K_n. For each set S of k vertices of K_n, let A_S be the event that the complete graph on S is monochromatic. Clearly, $\Pr(A_S) = 2^{1-\binom{k}{2}}$. It is obvious that each event A_s is mutually independent of all the events A_T but those that satisfy $|S \cap T| \geq 2$, since this is the only case in which the corresponding complete graphs share an edge. We can therefore apply Corollary 1.2 with $p = 2^{1-\binom{k}{2}}$ and $d = \binom{k}{2}\binom{n}{k-2}$ to conclude:

Proposition 3.1. *If $e(\binom{k}{2}\binom{n}{k-2}+1)\cdot 2^{1-\binom{k}{2}} < 1$, then $R(k,k) > n$.*

A short computation shows that this gives

$$R(k,k) > \frac{\sqrt{2}}{e}(1+o(1))k2^{k/2},$$

only a factor 2 improvement on the bound obtained by the straightforward probabilistic method. Although this minor improvement is somewhat disappointing, it is certainly not surprising; the local lemma is most powerful when the dependencies between events are rare, and this is not the case here. Indeed, there is a total number of $K = \binom{n}{k}$ events considered, and the maximum outdegree d in the dependency digraph is roughly $\binom{k}{2}\binom{n}{k-2}$. For large k and much larger n (which is the case of interest for us), we have $d > K^{1-O(1/k)}$, i.e., quite a lot of dependencies. On the other hand, if we consider small sets S, e.g., sets of size 3, we observe that out of the total $K = \binom{n}{3}$ of them, each shares an edge with only $3(n-3) \approx K^{1/3}$. This suggests that the local lemma may be much more significant in improving the off-diagonal Ramsey numbers $R(k,\ell)$, especially if one of the parameters, say ℓ, is small. Let us consider, for instance, following Spencer (1977), the Ramsey number $R(k,3)$. Here, of course, we have to apply the nonsymmetric form of the local lemma. Let us 2-color the edges of K_n randomly and independently, where each edge is colored blue with probability p. For each set of 3 vertices T, let A_T be the event that the triangle on T is blue. Similarly, for each set of k vertices S, let B_S be the event that the complete graph on S is red. Obviously, $\Pr(A_T) = p^3$ and $\Pr(B_S) = (1-p)^{\binom{k}{2}}$. Construct a dependency digraph for the events A_T and B_S by joining two vertices by edges (in both directions) iff the corresponding complete graphs share an edge. Clearly, each A_T-node of the dependency graph is adjacent to $3(n-3) < 3n$ $A_{T'}$-nodes and to at most $\binom{n}{k}$ $B_{S'}$-nodes. Similarly, each B_S-node is adjacent to at most $\binom{k}{2} \cdot n < k^2 n/2$ A_T-nodes and to at most $\binom{n}{k}$ $B_{S'}$-nodes. It follows from the general case of the local lemma (Lemma 1.1) that if we can find a $0 < p < 1$ and two real numbers $0 \le x < 1$ and $0 \le y < 1$ such that

$$p^3 \le x(1-x)^{3n}(1-y)^{\binom{n}{k}}$$

and

$$(1-p)^{\binom{k}{2}} \le y(1-x)^{k^2 n/2}(1-y)^{\binom{n}{k}},$$

then $R(k,3) > n$.

Our objective is to find the largest possible $k = k(n)$ for which there is such a choice of p, x and y. An elementary (but tedious) computation shows that the best choice is when $p = c_1 n^{-1/2}$, $k = c_2 n^{1/2} \log n$, $x = c_3/n^{3/2}$, and $y = c_4/\exp(n^{1/2}\log^2 n)$. This gives that $R(k,3) > c_5 k^2/\log^2 k$. A similar argument gives that $R(k,4) > k^{5/2+o(1)}$. In both cases, the amount of computation required is considerable. However, the hard work does pay; the bound $R(k,3) > c_5 k^2/\log^2 k$ matches a lower bound of Erdős proved in 1961 by a highly complicated probabilistic argument. The bound above for $R(k,4)$ is better than any bound for $R(k,4)$ known to be proven without the local lemma.

4. A GEOMETRIC RESULT

A family of open unit balls F in the 3-dimensional Euclidean space $\mathbb{R}^3$ is called a *k-fold covering* of $\mathbb{R}^3$ if any point $x \in \mathbb{R}^3$ belongs to at least k balls. In particular, a 1-fold covering is simply called a *covering*. A k-fold covering $\mathcal{F}$ is called *decomposable* if there is a partition of $\mathcal{F}$ into two pairwise disjoint families $\mathcal{F}_1$ and $\mathcal{F}_2$, each being a covering of $\mathbb{R}^3$. Mani-Levitska and Pach (1988) constructed, for any integer $k \geq 1$, a nondecomposable k-fold covering of $\mathbb{R}^3$ by open unit balls. On the other hand, they proved that any k-fold covering of $\mathbb{R}^3$ in which no point is covered by more than $c2^{k/3}$ balls is decomposable. This reveals a somewhat surprising phenomenon that it is more difficult to decompose coverings that cover some of the points of $\mathbb{R}^3$ too often, than to decompose coverings that cover every point about the same number of times. The exact statement of the Mani-Levitska–Pach theorem is the following.

Theorem 4.1. *Let $\mathcal{F} = \{B_i\}_{i \in I}$ be a k-fold covering of the 3-dimensional Euclidean space by open unit balls. Suppose, further, that no point of $\mathbb{R}^3$ is contained in more than t members of $\mathcal{F}$. If*

$$e \cdot \frac{t^3 2^{18}}{2^{k-1}} \leq 1,$$

then $\mathcal{F}$ is decomposable.

Proof. Let $\{C_j\}_{j \in J}$ be the connected components of the set obtained from $\mathbb{R}^3$ by deleting all the boundaries of the balls B_i in $\mathcal{F}$. Let $H = (V(H), E(H))$ be the (infinite) hypergraph defined as follows: The set of vertices of H, $V(H)$ is simply $\mathcal{F} = \{B_i\}_{i \in I}$. The set of edges of H is $E(H) = \{E_j\}_{j \in J}$, where $E_j = \{B_i : i \in I \text{ and } C_j \subseteq B_i\}$. Since $\mathcal{F}$ is a k-fold covering, each edge E_j of H contains at least k vertices. We claim that each edge of H intersects less than $t^3 2^{18}$ other edges of H. To prove this claim, fix an edge E_ℓ, corresponding to the connected component C_ℓ, where $\ell \in J$. Let E_j be an arbitrary edge of H, corresponding to the component C_j, that intersects E_ℓ. Then there is a ball B_i containing both C_ℓ and C_j. Therefore, any ball that contains C_j intersects B_i. It follows that all the unit balls that contain or touch a C_j, for some j that satisfies $E_j \cap E_\ell \neq \emptyset$, are contained in a ball B of radius 4. As no point of this ball is covered more than t times, we conclude, by a simple volume argument, that the total number of these unit balls is at most $t \cdot 4^3 = t \cdot 2^6$. It is not too difficult to check that m balls in $\mathbb{R}^3$ cut $\mathbb{R}^3$ into less than m^3 connected components, and since each of the above C_j is such a component, we have

$$\left|\{j : E_j \cap E_\ell \neq \emptyset\}\right| < (t \cdot 2^6)^3 = t^3 2^{18},$$

as claimed.

Consider, now, any finite subhypergraph L of H. Each edge of L has at least k vertices, and it intersects at most $d < t^3 2^{18}$ other edges of L. Since, by assumption, $e(d+1) \leq 2^{k-1}$, Theorem 2.1 (which is a simple corollary of the local lemma), implies that L is 2-colorable. This means that one can color the vertices of L blue and red so that no edge of L is monochromatic. Since this holds for any finite L, a compactness argument, analogous to the one used in the proof of Theorem 2.2, shows that H is 2-colorable. Given a 2-coloring of H with no monochromatic edges, we simply let $\mathcal{F}_1$ be the set of all blue balls, and $\mathcal{F}_2$ be the set of all red ones. Clearly, each $\mathcal{F}_i$ is a covering of $\mathbf{R}^3$, completing the proof of the theorem. ■

It is worth noting that Theorem 4.1 can be easily generalized to higher dimensions. We omit the detailed statement of this generalization.

5. THE LINEAR ARBORICITY OF GRAPHS

A *linear forest* is a forest (i.e., an acyclic simple graph) in which every connected component is a path. The *linear arboricity* $\text{la}(G)$ of a graph G is the minimum number of linear forests in G, whose union is the set of all edges of G. This notion was introduced by Harary as one of the covering invariants of graphs. The following conjecture, known as the *linear arboricity conjecture*, was raised in Akiyama, Exoo, and Harary (1981).

Conjecture 5.1 (The Linear Arboricity Conjecture). *The linear arboricity of every d-regular graph is* $\lceil (d+1)/2 \rceil$.

Notice that since every d-regular graph G on n vertices has $nd/2$ edges, and every linear forest in it has at most $n-1$ edges, the inequality

$$\text{la}(G) \geq \frac{nd}{2(n-1)} > \frac{d}{2}$$

is immediate. Since $\text{la}(G)$ is an integer, this gives $\text{la}(G) \geq \lceil (d+1)/2 \rceil$. The difficulty in Conjecture 5.1 lies in proving the converse inequality: $\text{la}(G) \leq \lceil (d+1)/2 \rceil$. Note also that since every graph G with maximum degree Δ is a subgraph of a Δ-regular graph (which may have more vertices, as well as more edges than G), the linear arboricity conjecture is equivalent to the statement that the linear arboricity of every graph G with maximum degree Δ is at most $\lceil (\Delta+1)/2 \rceil$.

Although this conjecture received a considerable amount of attention, the best general result concerning it, proved without any probabilistic arguments, is that $\text{la}(G) \leq \lceil 3\Delta/5 \rceil$ for even Δ and that $\text{la}(G) \leq \lceil (3\Delta+2)/5 \rceil$ for odd Δ. In this section, we prove that for every $\varepsilon > 0$ there is a $\Delta_0 = \Delta_0(\varepsilon)$ such that for every $\Delta \geq \Delta_0$, the linear arboricity of every graph with maximum degree Δ is less than $(1/2+\varepsilon)\Delta$. This result (with a somewhat more complicated proof) appears in Alon (1988) and its proof relies heavily on the local lemma. We note that this proof is more complicated than the other proofs given in this chapter, and requires certain preparations, some of which are of independent interest.

It is convenient to deduce the result for undirected graphs from its directed version. A *d-regular digraph* is a directed graph in which the indegree and the outdegree of every vertex is precisely d. A linear directed forest is a directed graph in which every connected component is a directed path. The *dilinear arboricity* $\text{dla}(G)$ of a directed graph G is the minimum number of linear directed forests in G whose union covers all edges of G. The directed version of the linear arboricity conjecture, first stated in Nakayama and Peroche (1987) is:

Conjecture 5.2. *For every d-regular digraph D,*

$$\text{dla}(D) = d + 1.$$

Note that since the edges of any (connected) undirected $2d$-regular graph G can be oriented along an Euler cycle, so that the resulting oriented digraph is d-regular, the validity of Conjecture 5.2 for d implies that of Conjecture 5.1 for $2d$.

It is easy to prove that any graph with n vertices and maximum degree d contains an independent set of size at least $n/(d+1)$. The following proposition shows that at the price of decreasing the size of such a set by a constant factor, we can guarantee that it has a certain structure.

Proposition 5.3. *Let $H = (V, E)$ be a graph with maximum degree d, and let $V = V_1 \cup V_2 \cup \cdots \cup V_r$ be a partition of V into r pairwise disjoint sets. Suppose each set V_i is of cardinality $|V_i| \geq 2ed$, where $e = 2.71828\ldots$ is the basis of the natural logarithm. Then there is an independent set of vertices $W \subseteq V$, that contains a vertex from each V_i.*

Proof. Clearly, we may assume that each set V_i is of cardinality precisely $g = \lceil 2ed \rceil$ (otherwise simply replace each V_i by a subset of cardinality g of it, and replace H by its induced subgraph on the union of these r new sets). Let us pick from each set V_i randomly and independently a single vertex according to a uniform distribution. Let W be the random set of the vertices picked. To complete the proof, we show that with positive probability, W is an independent set of vertices in H.

For each edge f of H, let A_f be the event that W contains both ends of f. Clearly, $\Pr(A_f) = 1/g^2$. Moreover, if the endpoints of f are in V_i and in V_j, then the event A_f is mutually independent of all the events corresponding to edges whose endpoints do not lie in $V_i \cup V_j$. Therefore there is a dependency digraph for the events in which the maximum degree is less than $2gd$, and since $e \cdot 2gd \cdot 1/g^2 = 2ed/g < 1$, we conclude, by Corollary 1.2, that with positive probability none of the events A_f holds. But this means that W is an independent set containing a vertex from each V_i, completing the proof. ■

Proposition 5.3 suffices to proves Conjecture 5.2 for digraphs with no short directed cycle. Recall that the directed girth of a digraph is the minimum length of a directed cycle in it.

Theorem 5.4. *Let $G = (U, F)$ be a d-regular digraph with directed girth $g \geq 4ed$. Then*

$$\text{dla}(G) = d + 1.$$

Proof. As is well known, F can be partitioned into d pairwise disjoint 1-regular spanning subgraphs $F_1, \ldots, F_d$ of G. (This is an easy consequence of the Hall–König theorem; let H be the bipartite graph whose two classes of vertices A and B are copies of U, in which $u \in A$ is joined to $v \in B$ iff $(u, v) \in F$. Since H is d-regular, its edges can be decomposed into d perfect matchings, which correspond to d 1-regular spanning subgraphs of G.) Each F_i is a union of vertex disjoint directed cycles $C_{i_1}, C_{i_2}, \ldots, C_{ir_i}$. Let $V_1, V_2, \ldots, V_r$ be the sets of *edges* of all the cycles $\{C_{ij} : 1 \leq i \leq d,\ 1 \leq j \leq r_i\}$. Clearly, $V_1, V_2, \ldots, V_r$ is a partition of the set F of all edges of G, and by the girth condition, $|V_i| \geq g \geq 4ed$ for all $1 \leq i \leq r$. Let H be the line graph of G, i.e., the graph whose set of vertices is the set F of edges of G in which two edges are adjacent iff they share a common vertex in G. Clearly, H is $2d - 2$ regular. As the cardinality of each V_i is at least $4ed \geq 2e(2d - 2)$, there is, by Proposition 5.3, an independent set of H containing a member from each V_i. But this means that there is a matching M in G, containing at least one edge from each cycle C_{ij} of the 1-factors $F_1, \ldots, F_d$. Therefore $M, F_1 \backslash M, F_2 \backslash M, \ldots, F_d \backslash M$ are $d + 1$-directed forests in G (one of which is a matching) that cover all its edges. Hence

$$\text{dla}(G) \leq d + 1.$$

As G has $|U| \cdot d$ edges and each directed linear forest can have at most $|U| - 1$ edges,

$$\text{dla}(G) \geq \frac{|U|d}{(|U| - 1)} > d.$$

Thus $\text{dla}(G) = d + 1$, completing the proof. ■

The last theorem shows that the assertion of Conjecture 5.2 holds for digraphs with sufficiently large (directed) girth. In order to deal with digraphs with small girth, we show that most of the edges of each regular digraph can be decomposed into a relatively small number of almost regular digraphs with high girth. To do this, we need the following statement, which is proved using the local lemma.

Lemma 5.5. *Let $G = (V, E)$ be a d-regular directed graph, where $d \geq 100$, and let p be an integer satisfying $10\sqrt{d} \leq p \leq 20\sqrt{d}$. Then there is a p-coloring of the vertices of G by the colors $0, 1, 2, \ldots, p - 1$ with the following property; for each vertex $v \in V$ and each color i, the numbers $N^+(v, i) = |\{u \in V;\ (v, u) \in E$*

and u is colored $i\}|$ and $N^-(v,i) = |\{u \in V : (u,v) \in E$ and u is colored $i\}|$ satisfy:

$$
\begin{aligned}
\left|N^+(v,i) - \frac{d}{p}\right| &\leq 3\sqrt{\frac{d}{p}}\sqrt{\log d}, \\
\left|N^-(v,i) - \frac{d}{p}\right| &\leq 3\sqrt{\frac{d}{p}}\sqrt{\log d}.
\end{aligned}
\tag{7}
$$

Proof. Let $f : V \to \{0,1,\dots,p-1\}$ be a random vertex coloring of V by p colors, where for each $v \in V$, $f(v) \in \{0,1,\dots,p-1\}$ is chosen according to a uniform distribution. For every vertex $v \in V$ and every color i, $0 \leq i < p$, let $A^+_{v,i}$ be the event that the number $N^+(v,i)$ of neighbors of v in G whose color is i does not satisfy inequality (7). Clearly, $N^+(v,i)$ is a binomial random variable with expectation d/p and standard deviation $\sqrt{(d/p)(1-1/p)} < \sqrt{d/p}$. Hence, by the standard estimates for binomial distribution given in Appendix A, for every $v \in V$ and $0 \leq i < p$,

$$\Pr(A^+_{v,i}) < \exp\left(-\frac{9\log d}{2}\right) < \frac{1}{d^4}.$$

Similarly, if $A^-_{v,i}$ is the event that the number $N^-(v,i)$ violates (7), then

$$\Pr(A^-_{v,i}) < \frac{1}{d^4}.$$

Clearly, each of the events $A^+_{v,i}$ or $A^-_{v,i}$ is mutually independent of all the events $A^+_{u,j}$ or $A^-_{u,j}$ for all vertices $u \in V$ that do not have a common neighbor with v in G. Therefore there is a dependency digraph for all our events with maximum degree $\leq (2d)^2 \cdot p$. Since $e \cdot (1/d^4)((2d)^2 p + 1) < 1$, Corollary 1.2 (i.e., the symmetric form of the local lemma), implies that with positive probability, no event $A^+_{v,i}$ or $A^-_{v,i}$ occurs. Hence there is a coloring f that satisfies (7) for all $v \in V$ and $0 \leq i < p$, completing the proof. ∎

We are now ready to deal with general regular digraphs. Let $G = (V,E)$ be an arbitrary d-regular digraph. Throughout the argument we assume, whenever it is needed, that d is sufficiently large. Let p be a prime satisfying $10d^{1/2} \leq p \leq 20d^{1/2}$ (it is well known that for every n there is a prime between n and $2n$). By Lemma 5.5 there is a vertex coloring $f : V \to \{0,1,\dots,p-1\}$ satisfying (7). For each i, $0 \leq i < p$, let $G_i = (V,E_i)$ be the spanning subdigraph of G defined by $E_i = \{(u,v) \in E : f(v) \equiv (f(u)+i) \bmod p\}$. By inequality (7) the maximum indegree Δ^-_i and the maximum outdegree Δ^+_i in each G_i is at most $d/p + 3\sqrt{d/p}\sqrt{\log d}$. Moreover, for each $i > 0$, the length of every directed cycle in G_i is divisible by p. Thus the directed girth g_i of G_i is at least p. Since each G_i can be completed, by adding vertices and edges, to a Δ_i-regular digraph with the same girth g_i and with $\Delta_i = \max(\Delta_i^+, \Delta_i^-)$, and since $g_i > 4e\Delta_i$ (for all sufficiently large d), we conclude, by Theorem 5.4, that

$$\mathrm{dla}(G_i) \le \Delta_i + 1 \le \frac{d}{p} + 3\sqrt{\frac{d}{p}}\sqrt{\log d} + 1$$

for all $1 \le i < p$. For G_0, we only apply the trivial inequality

$$\mathrm{dla}(G_0) \le 2\Delta_0 \le 2\frac{d}{p} + 6\sqrt{\frac{d}{p}}\sqrt{\log d}$$

obtained by, e.g., embedding G_0 as a subgraph of a Δ_0-regular graph, splitting the edges of this graph into Δ_0 1-regular spanning subgraphs, and breaking each of these 1-regular spanning subgraphs into two linear directed forests. The last two inequalities, together with the fact that $10\sqrt{d} \le p \le 20\sqrt{d}$, imply

$$\mathrm{dla}(G) \le d + \frac{d}{p} + 3\sqrt{pd}\sqrt{\log d} + 3\sqrt{\frac{d}{p}}\sqrt{\log d} + p - 1 \le d + c \cdot d^{3/4}(\log d)^{1/2}.$$

We have thus proved:

Theorem 5.6. *There is an absolute constant $c > 0$ such that for every d-regular digraph G,*

$$\mathrm{dla}(G) \le d + cd^{3/4}(\log d)^{1/2}.$$

We note that by being a little more careful, we can improve the error term to $c'd^{2/3}(\log d)^{1/3}$. Since the edges of any undirected $d = 2f$-regular graph can be oriented so that the resulting digraph is f-regular, and since any $(2f - 1)$-regular undirected graph is a subgraph of a $2f$-regular graph, the last theorem implies:

Theorem 5.7. *There is an absolute constant $c > 0$ such that for every undirected d-regular graph G,*

$$\mathrm{la}(G) \le \frac{d}{2} + cd^{3/4}(\log d)^{1/2}.$$

6. LATIN TRANSVERSALS

Following the proof of the local lemma, we noted that the mutual independency assumption in this lemma can be replaced by the weaker assumption that the conditional probability of each event, given the mutual nonoccurrence of an arbitrary set of events, each nonadjacent to it in the dependency digraph, is sufficiently small. In this section, we describe an application, from Erdős and Spencer (1991), of this modified version of the lemma. Let $A = (a_{ij})$ be an n by n matrix with, say, integer entries. A permutation π is called a *Latin transversal* (of A) if the entries $a_{i\pi(i)}$ $(1 \le i \le n)$ are all distinct.

Theorem 6.1. *Suppose $k \le (n-1)/(4e)$ and suppose that no integer appears in more than k entries of A. Then A has a Latin transversal.*

Proof. Let π be a random permutation of $\{1,2,\ldots,n\}$, chosen according to a uniform distribution among all possible $n!$ permutations. Denote by T the set of all ordered fourtuples (i,j,i',j') satisfying $i < i'$, $j \ne j'$ and $a_{ij} = a_{i'j'}$. For each $(i,j,i',j') \in T$, let $A_{iji'j'}$ denote the event that $\pi(i) = j$ and $\pi(i') = j'$. The existence of a Latin transversal is equivalent to the statement that with positive probability none of these events hold. Let us define a symmetric digraph (i.e., a graph) G on the vertex set T by making (i,j,i',j') adjacent to (p,q,p',q') if and only if $\{i,i'\} \cap \{p,p'\} \ne \emptyset$ or $\{j,j'\} \cap \{q,q'\} \ne \emptyset$. Thus these two fourtuples are not adjacent iff the four cells (i,j), (i',j'), (p,q) and (p',q') occupy four distinct rows and columns of A. The maximum degree of G is less than $4nk$; indeed, for a given $(i,j,i',j') \in T$, there are $4n$ choices of (p,q) with either $p \in \{i,i'\}$ or $q \in \{j,j'\}$, and for each of these choices of (p,q) there are less than k choices for $(p',q') \ne (p,q)$ with $a_{pq} = a_{p'q'}$. Since $e \cdot 4nk \cdot 1/[n(n-1)] \le 1$, the desired result follows from the above-mentioned strengthening of the symmetric version of the local lemma if we can show that

$$\Pr\left(A_{iji'j'} \,\middle|\, \bigwedge_S \overline{A}_{pqp'q'}\right) \le \frac{1}{n(n-1)} \tag{8}$$

for any $(i,j,i',j') \in T$ and any set S of members of T that are nonadjacent in G to (i,j,i',j'). By symmetry, we may assume that $i = j = 1$, $i' = j' = 2$, and that hence none of the p's nor q's are either 1 or 2. Let us call a permutation π *good* if it satisfies $\bigwedge_S \overline{A}_{pqp'q'}$, and let S_{ij} denote the set of all good permutations π satisfying $\pi(1) = i$ and $\pi(2) = j$. We claim that $|S_{12}| \le |S_{ij}|$ for all $i \ne j$. Indeed, suppose first that $i,j > 2$. For each good $\pi \in S_{12}$, define a permutation π^* as follows. Suppose $\pi(x) = i$, $\pi(y) = j$. Then define $\pi^*(1) = i$, $\pi^*(2) = j$, $\pi^*(x) = 1$, $\pi^*(y) = 2$ and $\pi^*(t) = \pi(t)$ for all $t \ne 1,2,x,y$. One can easily check that π^* is good, since the cells $(1,i),(2,j),(x,1),(y,2)$ are not part of any $(p,q,p',q') \in S$. Thus $\pi^* \in S_{ij}$, and since the mapping $\pi \to \pi^*$ is injective, $|S_{12}| \le |S_{ij}|$, as claimed. Similarly, one can define injective mappings showing that $|S_{12}| \le |S_{ij}|$ even when $\{i,j\} \cap \{1,2\} \ne \emptyset$. It follows that

$$\Pr\left(A_{1122} \wedge \bigwedge_S \overline{A}_{pqp'q'}\right) \le \Pr\left(A_{1i2j} \wedge \bigwedge_S \overline{A}_{pqp'q'}\right)$$

for all $i \ne j$ and hence that

$$\Pr\left(A_{1122} \,\middle|\, \bigwedge_S \overline{A}_{pqp'q'}\right) \le 1/n(n-1).$$

By symmetry, this implies (8) and completes the proof. ■

7. THE ALGORITHMIC ASPECT

When the probabilistic method is applied to prove that a certain event holds with high probability, it often supplies an efficient deterministic, or at least randomized, algorithm for the corresponding problem.

By applying the local lemma, we often manage to prove that a given event holds with positive probability, although this probability may be exponentially small in the dimensions of the problem. Consequently, it is not clear if any of these proofs can provide a polynomial algorithm for the corresponding algorithmic problems. For several years, there has been no known method of converting the proofs of any of the examples discussed in this chapter into an efficient algorithm. Very recently, J. Beck (1991) found such a method that works for some of these examples, with a little loss in the constants. Beck demonstrated his method by considering the problem of hypergraph 2-coloring. For simplicity, we only describe here the case of fixed edge-size in which each edge intersects a fixed number of other edges.

Let n, d be fixed positive integers. By the (n,d)-problem, we mean the following: Given sets $A_1, \ldots, A_N \subseteq \Omega$ with all $|A_i| = n$, such that no set A_i intersects more than d other sets A_j, find a 2-coloring of Ω so that no A_i is monochromatic. When $e(d+1) < 2^{n-1}$, Theorem 2.1 assures us that this problem always does have a solution. Can we find the coloring in polynomial (in N for fixed n,d) time? J. Beck has given an affirmative answer under somewhat more restrictive assumptions. We assume Ω is of the form $\Omega = \{1, \ldots, m\}$, $m \leq Nn$, and the initial data structure consists of a list of the elements of the sets A_i and a list giving for each element j those i for which $j \in A_i$. We let G denote the dependency graph with vertices the sets A_i and A_i, A_j adjacent if they overlap.

Theorem 7.1. *Let n,d be such that, setting $D = d(d-1)^3$, there exists a decomposition $n = n_1 + n_2 + n_3$ with*

$$16D(1+d) < 2^{n_1},$$

$$16D(1+d) < 2^{n_2},$$

$$2e(1+d) < 2^{n_3}.$$

Then there is a randomized algorithm with expected running time $O(N(\ln N)^c)$ for the (n,d) problem, where c is a constant (depending only on n and d).

For $\epsilon < 1/11$, fixed, we note that the above conditions are satisfied, for n sufficiently large, when $d < 2^{n\epsilon}$ by taking $n_1 = n_2 \sim 5n/11$ and $n_3 \sim n/11$. We emphasize again that the algorithmic analysis here is for *fixed* n,d and N approaching infinity, although the argument can be extended to the nonfixed case as well.

Beck (1991) has given a deterministic algorithm for the (n,d)-problem. The randomized algorithm we give may be derandomized using the techniques of

Chapter 15. The running time remains polynomial but seemingly no longer $N^{1+o(1)}$. Moreover, the algorithm can even be parallelized using some of the techniques in Chapter 15 together with a certain modification in the algorithm.

The First Pass. During this pass, points will be either red, blue, uncolored, or saved. We move through the points $j \in \Omega$ sequentially, coloring them red or blue at random, flipping a fair coin. After each j is colored, we check all $A_i \ni j$. If A_i now has n_1 points in one color and no points in the other color, we call A_i *dangerous*. All uncolored $k \in A_i$ are now considered saved. When saved points k are reached in the sequential coloring, they are not colored but simply skipped over. At the conclusion of the first pass, points are red, blue, or saved. We say a set A_i *survives* if it does not have both red and blue points. Let $S \subseteq G$ denote the (random) set of survived sets.

Claim 7.2. *Almost surely all components C of $G|_S$ have size $O(\ln N)$.*

Proof. An $A_i \in S$ may be dangerous or, possibly, many of its points were saved because neighboring (in G) sets were dangerous. The probability of a particular A_i becoming dangerous is at most 2^{1-n_1}, since for this to occur the first n_1 coin flips determining colors of $j \in A_i$ must come up the same. (We only have inequality, since in addition n_1 points of A_i must be reached while still uncolored.) Let V be an independent set in G, i.e., the $A_i \in V$ are mutually disjoint. Then the probability that all $A_i \in V$ become dangerous is at most $(2^{1-n_1})^{|V|}$, as the coin flips involve disjoint sets. Now let $V \subseteq G$ be such that all distances between the $A_i \in V$ are at least 4, distance being the length of the shortest path in G. We claim that

$$\Pr[V \subseteq S] \leq (d+1)^{|V|}(2^{1-n_1})^{|V|}.$$

This is because for each $A_i \in V$ there are $d+1$ choices for a dangerous neighbor $A_{i'}$, giving $(d+1)^{|V|}$ choices for the $A_{i'}$. As the A_i are at least 4 apart, the $A_{i'}$ cannot be adjacent and so the probability that they are all dangerous is at most $(2^{1-n_1})^{|V|}$, as claimed.

Call $T \subseteq G$ a *4-tree* if the $A_i \in T$ are such that all their mutual distances in G are at least 4 and so that, drawing an arc between $A_i, A_j \in T$ if their distance is precisely 4, the resulting graph is connected. We first bound the number of 4-trees of size u. The "distance-four" graph defined on T must contain a tree. There are less than 4^j trees (up to isomorphism) on j vertices; now fix one. We can label the tree $1,\dots,u$, so that each $j > 1$ is adjacent to some $i < j$. Now consider the number of $(A^1,\dots,A^u)$ whose distance-four graph corresponds to this tree. There are N choices for A^1. Having chosen A^i for all $i < j$, the set A^j must be at distance 4 from A^i in G and there are at most D such points. Hence the number of 4-trees of size u is at most $4^u N D^{u-1} < N(4D)^u$. For any particular 4-tree T, we have already that

$$\Pr[T \subseteq S] \leq [(d+1)2^{1-n_1}]^u.$$

Therefore the expected number of 4-trees $T \subseteq S$ is at most

$$N\left[8D(d+1)2^{-n_1}\right]^u.$$

As the bracketed term is less than 1/2 by assumption, for $u = c_1 \ln N$ this term is $o(1)$. Thus almost surely $G|_S$ will contain no 4-tree of size bigger than $c_1 \ln N$. We actually want to bound the size of the components C of $G|_S$. A maximal 4-tree T in a component C must have the property that every $A_i \in C$ lies within 3 of an $A_j \in T$. There are less than d^3 (a constant) A_i within 3 of any given A_j, so that $c_1 \ln N \geq |T| \geq |C| d^{-3}$ and so (since d is a constant)

$$|C| \leq c_2 \ln N,$$

proving the claim. ■

If the first pass leaves components of size larger than $c_2 \ln N$, we simply repeat the entire procedure. In expected *linear* time, the first pass is successful. The points that are red or blue are now fixed. The sets A_i with both red and blue points can now be ignored. For each surviving A_i, fix a subset B_i of $n - n_1$ saved points. It now suffices to color the saved points so that no B_i is monochromatic. The B_i split into components of size $O(\ln N)$ and it suffices to color each component separately. On the second pass, we apply the method of the first pass to each component of the B_i. Now we call a set B_i dangerous if it receives n_2 points of one color and none of the other. The second pass takes expected time $O(M)$ to color a component of size M, hence an expected time $O(N)$ to color all the components. (For success, we require that a component of size M is broken into components of size at most $c_2 \ln M$. To avoid trivialities, if $M < \ln \ln N$, we skip the second pass for the corresponding component.) At the end of the second pass (still in linear time!), there is a family of twice surviving sets $C_i \subset B_i \subset A_i$ of size n_3, the largest component of which has size $O(\ln \ln N)$.

We still need to color these $O(N)$ components of sets of size n_3, each component of size $O(\ln \ln N)$. By the local lemma (or directly by Theorem 2.1), each of these components can be 2-colored. *We now find the 2-coloring by brute force!* Examining all 2-colorings of a component of size M takes time $O(M 2^M)$, which is $O((\ln N)^c)$ in our case. Doing this for all components takes time $O(N(\ln N)^c)$. This completes the coloring. ■

We note that with slightly more restrictions on n, d, a third pass could be made and then the total time would be $O(N(\ln \ln N)^c)$. We note also that a similar technique can be applied for converting several other applications of the local lemma into efficient algorithms.

The Probabilistic Lens: Directed Cycles

Let $D = (V, E)$ be a simple directed graph with minimum outdegree δ and maximum indegree Δ.

Theorem (Alon and Linial [1989]). *If* $e(\Delta\delta + 1)(1 - 1/k)^{\delta} < 1$, *then* D *contains a (directed, simple) cycle of length* $0(\bmod k)$.

Proof. Clearly, we may assume that every outdegree is precisely δ, since otherwise we can consider a subgraph of D with this property.

Let $f : V \to \{0, 1, \ldots, k-1\}$ be a random coloring of V, obtained by choosing, for each $v \in V$, $f(v) \in \{0, \ldots, k-1\}$ independently, according to a uniform distribution. For each $v \in V$, let A_v denote the event that there is no $u \in V$, with $(v, u) \in E$ and $f(u) \equiv (f(v) + 1)(\bmod k)$. Clearly, $\Pr(A_v) = (1 - 1/k)^{\delta}$. One can easily check that each event A_v is mutually independent of all the events A_u but those satisfying

$$N^+(v) \cap (u \cup N^+(u)) \neq \emptyset,$$

where here $N^+(v) = \{w \in V : (v, w) \in E\}$. The number of such u's is at most $\Delta\delta$ and hence, by our assumption and by the local lemma (Corollary 1.2 in Chapter 5), $\Pr(\bigwedge_{v \in V} \overline{A_v}) > 0$. Therefore there is an $f : V \to \{0, 1, \ldots, k-1\}$ such that for every $v \in V$ there is a $u \in V$ with

$$(v, u) \in E \qquad \text{and} \qquad f(u) \equiv (f(v) + 1)(\bmod k). \qquad (*)$$

Starting at an arbitrary $v = v_0 \in V$ and applying $(*)$ repeatedly, we obtain a sequence $v_0, v_1, v_2, \ldots$ of vertices of D so that $(v_i, v_{i+1}) \in E$ and $f(v_{i+1}) \equiv (f(v_i) + 1)(\bmod k)$ for all $i \geq 0$. Let j be the minimum integer so that there is an $\ell < j$ with $v_\ell = v_j$. The cycle $v_\ell v_{\ell+1} v_{\ell+2} \cdots v_j = v_\ell$ is a directed simple cycle of D whose length is divisible by k. ∎

6

Correlation Inequalities

Let $G = (V, E)$ be a random graph on the set of vertices $V = \{1, 2, \ldots, n\}$ generated by choosing, for each $i, j \in V$, $i \neq j$ independently, the pair $\{i, j\}$ to be an edge with probability p, where $0 < p < 1$. Let H be the event that G is Hamiltonian and let P be the event that G is planar. Suppose one wants to compare the two quantities $\Pr(P \wedge H)$ and $\Pr(P) \cdot \Pr(H)$. Intuitively, knowing that G is Hamiltonian suggests that it has many edges and hence seems to indicate that G is less likely to be planar. Therefore it seems natural to expect that $\Pr(P|H) \leq \Pr(P)$, implying

$$\Pr(P \wedge H) \leq \Pr(H) \cdot \Pr(P).$$

This inequality, which is, indeed, correct, is a special case of the FKG *inequality* of Fortuin, Kasteleyn, and Ginibre (1971). In this chapter, we present the proof of this inequality and several related results, which deal with the correlation between certain events in probability spaces. The proofs of all these results are rather simple, and still they supply many interesting consequences. The first inequality of this type is due to Harris (1960). A result closer to the ones considered here is a lemma of Kleitman (1966b), stating that if $\mathcal{A}$ and $\mathcal{B}$ are two *monotone decreasing* families of subsets of $\{1, 2, \ldots, n\}$ (i.e., $A \in \mathcal{A}$ and $A' \subseteq A \Rightarrow A' \in \mathcal{A}$ and, similarly $B \in \mathcal{B}$ and $B' \subseteq B \Rightarrow B' \in \mathcal{B}$), then

$$|\mathcal{A} \cap \mathcal{B}| \cdot 2^n \geq |\mathcal{A}| \cdot |\mathcal{B}|.$$

This lemma was followed by many extensions and generalizations until 1978, when Ahlswede and Daykin obtained a very general result, which implies all these extensions. In the next section, we present this result and its proof. Some of its many applications are discussed in the rest of the chapter.

1. THE FOUR FUNCTIONS THEOREM OF AHLSWEDE AND DAYKIN

Suppose $n \geq 1$ and put $N = \{1, 2, \ldots, n\}$. Let $P(N)$ denote the set of all subsets of N, and let $\mathbf{R}^+$ denote the set of nonnegative real numbers. For a function $\varphi : P(N) \to \mathbf{R}^+$ and for a family $\mathcal{A}$ of subsets of N, denote $\varphi(\mathcal{A}) = \sum_{A \in \mathcal{A}} \varphi(A)$. For two families $\mathcal{A}$ and $\mathcal{B}$ of subsets of N, define $\mathcal{A} \cup \mathcal{B} = \{A \cup B : A \in \mathcal{A},\ B \in \mathcal{B}\}$ and $\mathcal{A} \cap \mathcal{B} = \{A \cap B : A \in \mathcal{A},\ B \in \mathcal{B}\}$.

Theorem 1.1 (The Four Functions Theorem). *Let $\alpha, \beta, \gamma, \delta : P(N) \to \mathrm{R}^+$ be four functions from the set of all subsets of N to the nonnegative reals. If, for every two subsets $A, B \subseteq N$, the inequality*

$$\alpha(A)\beta(B) \leq \gamma(A \cup B)\delta(A \cap B) \tag{1}$$

holds, then, for every two families of subsets $\mathcal{A}, \mathcal{B} \subseteq P(N)$,

$$\alpha(\mathcal{A})\beta(\mathcal{B}) \leq \gamma(\mathcal{A} \cup \mathcal{B})\delta(\mathcal{A} \cap \mathcal{B}). \tag{2}$$

Proof. Observe, first, that we may modify the four functions $\alpha, \beta, \gamma, \delta$ by defining $\alpha(A) = 0$ for all $A \notin \mathcal{A}$, $\beta(B) = 0$ for all $B \notin \mathcal{B}$, $\gamma(C) = 0$ for all $C \notin \mathcal{A} \cup \mathcal{B}$, and $\delta(D) = 0$ for all $D \notin \mathcal{A} \cap \mathcal{B}$. Clearly, (1) still holds for the modified functions and in inequality (2) we may assume now that $\mathcal{A} = \mathcal{B} = \mathcal{A} \cup \mathcal{B} = \mathcal{A} \cap \mathcal{B} = P(N)$.

To prove this inequality, we apply induction on n. The only step that requires some computation is $n = 1$. In this case, $P(N) = \{\phi, N\}$. For each function $\varphi \in \{\alpha, \beta, \gamma, \delta\}$, define $\varphi_0 = \varphi(\phi)$ and $\varphi_1 = \varphi(N)$. By (1), we have

$$\begin{aligned} \alpha_0\beta_0 &\leq \gamma_0\delta_0, \\ \alpha_0\beta_1 &\leq \gamma_1\delta_0, \\ \alpha_1\beta_0 &\leq \gamma_1\delta_0, \\ \alpha_1\beta_1 &\leq \gamma_1\delta_1. \end{aligned} \tag{3}$$

By the above paragraph we only have to prove inequality (2), where $\mathcal{A} = \mathcal{B} = P(N)$, i.e., to prove that

$$(\alpha_0 + \alpha_1)(\beta_0 + \beta_1) \leq (\gamma_0 + \gamma_1)(\delta_0 + \delta_1). \tag{4}$$

If either $\gamma_1 = 0$ or $\delta_0 = 0$, this follows immediately from (3). Otherwise, by (3), $\gamma_0 \geq \alpha_0\beta_0/\delta_0$ and $\delta_1 \geq \alpha_1\beta_1/\gamma_1$. It thus suffices to show that

$$\left(\frac{\alpha_0\beta_0}{\delta_0} + \gamma_1\right)\left(\delta_0 + \frac{\alpha_1\beta_1}{\gamma_1}\right) \geq (\alpha_0 + \alpha_1)(\beta_0 + \beta_1),$$

or, equivalently, that

$$(\alpha_0\beta_0 + \gamma_1\delta_0)(\delta_0\gamma_1 + \alpha_1\beta_1) \geq (\alpha_0 + \alpha_1)(\beta_0 + \beta_1)\delta_0\gamma_1.$$

The last inequality is equivalent to

$$(\gamma_1\delta_0 - \alpha_0\beta_1)(\gamma_1\delta_0 - \alpha_1\beta_0) \geq 0,$$

which follows from (3), as both factors in the left-hand side are nonnegative. This completes the proof for $n = 1$.

Suppose, now, that the theorem holds for $n - 1$ and let us prove it for n, $(n \geq 2)$. Put $N' = N \backslash \{n\}$ and define for each $\varphi \in \{\alpha, \beta, \gamma, \delta\}$ and each $A \subseteq$

N', $\varphi'(A) = \varphi(A) + \varphi(A \cup \{n\})$. Clearly, for each function $\varphi \in \{\alpha,\beta,\gamma,\delta\}$ $\varphi'(P(N')) = \varphi(P(N))$. Therefore the desired inequality (2) would follow from applying the induction hypothesis to the functions $\alpha',\beta',\gamma',\delta' : P(N') \to \mathbb{R}^+$. However, in order to apply this hypothesis, we have to check that these new functions satisfy the assumption of Theorem 1.1 on N', i.e., that for every $A',B' \subseteq N'$,

$$\alpha'(A')\beta'(B') \leq \gamma'(A' \cup B')\delta'(A' \cap B'). \tag{5}$$

Not surprisingly, this last inequality follows easily from the case $n = 1$, which we have already proved. Indeed, let T be a 1-element set and define $\overline{\alpha}(\phi) = \alpha(A')$, $\overline{\alpha}(T) = \alpha(A' \cup \{n\})$, $\overline{\beta}(\phi) = \beta(B')$, $\overline{\beta}(T) = \beta(B' \cup \{n\})$, $\overline{\gamma}(\phi) = \gamma(A' \cup B')$, $\overline{\gamma}(T) = \gamma(A' \cup B' \cup \{n\})$, and $\overline{\delta}(\phi) = \delta(A' \cap B')$, $\overline{\delta}(T) = \delta((A' \cap B') \cup \{n\})$. By the assumption (1), $\overline{\alpha}(S)\overline{\beta}(R) \leq \overline{\gamma}(S \cup R)\overline{\delta}(S \cap R)$ for all $S,R \subseteq T$ and hence, by the case $n = 1$ already proven, $\alpha'(A')\beta'(B') = \overline{\alpha}(P(T))\overline{\beta}(P(T)) \leq \overline{\gamma}(P(T))\overline{\delta}(P(T)) = \gamma'(A' \cup B')\delta'(A' \cap B')$, which is the desired inequality (5). Therefore inequality (2) holds, completing the proof. ■

The Ahlswede–Daykin theorem can be extended to arbitrary finite distributive lattices. A *lattice* is a partially ordered set in which every two elements x and y have a unique minimal upper bound, denoted by $x \vee y$ and called the *join* of x and y, and a unique maximal lower bound, denoted by $x \wedge y$ and called the *meet* of x and y. A lattice L is *distributive* if for all $x,y,z \in L$,

$$x \wedge (y \vee z) = (x \wedge y) \vee (x \wedge z)$$

or, equivalently if for all $x,y,z \in L$,

$$x \vee (y \wedge z) = (x \vee y) \wedge (x \vee z).$$

For two sets $X,Y \subseteq L$, define

$$X \vee Y = \{x \vee y : x \in X,\ y \in Y\}.$$

and

$$X \wedge Y = \{x \wedge y : x \in X,\ y \in Y\}.$$

Any subset L of $P(N)$, where $N = \{1,2,\ldots,n\}$, ordered by inclusion, which is closed under the union and intersection operations is a distributive lattice. Here, the join of two members $A,B \in L$, is simply their union $A \cup B$ and their meet is the intersection $A \cap B$. It is somewhat more surprising (but easy to check) that every finite distributive lattice L is isomorphic to a sublattice of $P(\{1,2,\ldots,n\})$ for some n. (To see this, call an element $x \in L$ *join-irreducible* if whenever $x = y \vee z$, then either $x = y$ or $x = z$. Let $x_1,x_2,\ldots,x_n$ be the set of all join-irreducible elements in L and associate each element $x \in L$ with the set $A = A(x) \subseteq N$, where $x = \bigvee_{i \in A} x_i$ and $\{x_i : i \in A\}$ are all the join-irreducibles y satisfying $y \leq x$. The mapping $x \to A(x)$ is the desired isomorphism.) This fact enables us to generalize Theorem 1.1 to arbitrary finite distributive lattices as follows.

Corollary 1.2. *Let L be a finite distributive lattice and let α, β, γ and δ be four functions from L to* $\mathbf{R}^+$. *If*

$$\alpha(x)\beta(y) \leq \gamma(x \vee y)\delta(x \wedge y)$$

for all $x, y \in L$, then for every $X, Y \subseteq L$,

$$\alpha(X)\beta(Y) \leq \gamma(X \vee Y)\delta(X \wedge Y).$$

The simplest case in the last corollary is the case where all the four functions α, β, γ, and δ are identically 1, stated below.

Corollary 1.3. *Let L be a finite distributive lattice and suppose $X, Y \subseteq L$. Then*

$$|X| \cdot |Y| \leq |X \vee Y| \cdot |X \wedge Y|.$$

We close this section by presenting a very simple consequence of the last corollary, first proved by Marica and Schonheim (1969).

Corollary 1.4. *Let X be a family of subsets of a finite set N and define*

$$X \backslash X = \{F \backslash F' : F,\ F' \in X\}.$$

Then $|X \backslash X| \geq |X|$.

Proof. Let L be the distributive lattice of all subsets of N. By applying Corollary 1.3 to X and $Y = \{N \backslash F : F \in X\}$, we obtain

$$|X|^2 = |X| \cdot |Y| \leq |X \cup Y| \cdot |X \cap Y| = |X \backslash X|^2.$$

The desired result follows. ∎

2. THE FKG INEQUALITY

A function $\mu : L \to \mathbf{R}^+$, where L is a finite distributive lattice, is called *log-supermodular* if

$$\mu(x)\mu(y) \leq \mu(x \vee y)\mu(x \wedge y)$$

for all $x, y \in L$. A function $f : L \to \mathbf{R}^+$ is *increasing* if $f(x) \leq f(y)$ whenever $x \leq y$ and is *decreasing* if $f(x) \geq f(y)$ whenever $x \leq y$.

Motivated by a problem from statistical mechanics, Fortuin, Kasteleyn, and Ginibre (1971) proved the following useful inequality, which has become known as the FKG inequality.

Theorem 2.1 (The FKG Inequality). *Let L be a finite distributive lattice and let $\mu : L \to \mathbf{R}^+$ be a log-supermodular function. Then, for any two increasing functions $f, g : L \to \mathbf{R}^+$, we have*

$$\left(\sum_{x\in L}\mu(x)f(x)\right)\cdot\left(\sum_{x\in L}\mu(x)g(x)\right)\leq\left(\sum_{x\in L}\mu(x)f(x)g(x)\right)\cdot\left(\sum_{x\in L}\mu(x)\right). \tag{6}$$

Proof. Define four functions $\alpha, \beta, \gamma, \delta : L \to \mathbf{R}^+$ as follows. For each $x \in L$,

$$\alpha(x) = \mu(x)f(x), \qquad \beta(x) = \mu(x)g(x),$$
$$\gamma(x) = \mu(x)f(x)g(x), \qquad \delta(x) = \mu(x).$$

We claim that these functions satisfy the hypothesis of the Ahlswede–Daykin theorem, stated in Corollary 1.2. Indeed, if $x, y \in L$, then, by the supermodularity of μ and since f and g are increasing,

$$\alpha(x)\beta(y) = \mu(x)f(x)\mu(y)g(y) \leq \mu(x \vee y)f(x)g(y)\mu(x \wedge y)$$
$$\leq \mu(x \vee y)f(x \vee y)g(x \vee y)\mu(x \wedge y) = \gamma(x \vee y)\delta(x \wedge y).$$

Therefore by Corollary 1.2 (with $X = Y = L$),

$$\alpha(L)\beta(L) \leq \gamma(L)\delta(L),$$

which is the desired result. ∎

Note that the conclusion of Theorem 2.1 holds also if both f and g are decreasing (simply interchange γ and δ in the proof). In case f is increasing and g is decreasing (or vice versa), the opposite inequality holds:

$$\left(\sum_{x\in L}\mu(x)f(x)\right)\left(\sum_{x\in L}\mu(x)g(x)\right)\geq\left(\sum_{x\in L}\mu(x)f(x)g(x)\right)\left(\sum_{x\in L}\mu(x)\right).$$

To prove it, simply apply Theorem 2.1 to the two increasing functions $f(x)$ and $k - g(x)$, where k is the constant $\max_{x\in L} g(x)$. (This constant is needed to guarantee that $k - g(x) \geq 0$ for all $x \in L$.)

It is helpful to view μ as a measure on L. Assuming μ is not identically 0, we can define, for any function $f : L \to \mathbf{R}^+$, its expectation

$$\langle f \rangle = \frac{\sum_{x\in L} f(x)\mu(x)}{\sum_{x\in L}\mu(x)}.$$

With this notation, the FKG inequality asserts that if μ is log-supermodular and $f, g : L \to \mathbf{R}^+$ are both increasing or both decreasing, then $\langle fg \rangle \geq \langle f \rangle \cdot \langle g \rangle$.

Similarly, if f is increasing and g is decreasing (or vice versa), then $\langle fg \rangle \leq \langle f \rangle \langle g \rangle$.

This formulation demonstrates clearly the probabilistic nature of the inequality, some of whose many interesting consequences are presented in the rest of this chapter.

3. MONOTONE PROPERTIES

Recall that a family $\mathcal{A}$ of subsets of $N = \{1,2,\ldots,n\}$ is *monotone decreasing* if $A \in \mathcal{A}$ and $A' \subseteq A \Rightarrow A' \in \mathcal{A}$. Similarly, it is *monotone increasing* if $A \in \mathcal{A}$ and $A \subseteq A' \Rightarrow A' \in \mathcal{A}$. By considering the power set $P(N)$ as a symmetric probability space, one naturally defines the *probability* of $\mathcal{A}$ by

$$\Pr(\mathcal{A}) = \frac{|\mathcal{A}|}{2^n}.$$

Thus $\Pr(\mathcal{A})$ is simply the probability that a randomly chosen subset of N lies in $\mathcal{A}$.

Kleitman's lemma (1966b), which was the starting point of all the correlation inequalities considered in this chapter, is the following.

Proposition 3.1. *Let $\mathcal{A}$ and $\mathcal{B}$ be two monotone increasing families of subsets of $N = \{1,2,\ldots,n\}$ and let $\mathcal{C}$ and $\mathcal{D}$ be two monotone decreasing families of subsets of N. Then*

$$\Pr(\mathcal{A}\cap\mathcal{B}) \geq \Pr(\mathcal{A})\cdot\Pr(\mathcal{B}),$$

$$\Pr(\mathcal{C}\cap\mathcal{D}) \geq \Pr(\mathcal{C})\cdot\Pr(\mathcal{D}),$$

$$\Pr(\mathcal{A}\cap\mathcal{C}) \leq \Pr(\mathcal{A})\cdot\Pr(\mathcal{C}).$$

In terms of cardinalities, this can be read as follows:

$$2^n|\mathcal{A}\cap\mathcal{B}| \geq |\mathcal{A}|\cdot|\mathcal{B}|,$$

$$2^n|\mathcal{C}\cap\mathcal{D}| \geq |\mathcal{C}|\cdot|\mathcal{D}|,$$

$$2^n|\mathcal{A}\cap\mathcal{C}| \leq |\mathcal{A}|\cdot|\mathcal{C}|.$$

Proof. Let $f : P(N) \to \mathbf{R}^+$ be the characteristic function of $\mathcal{A}$, i.e., $f(A) = 0$ if $A \notin \mathcal{A}$ and $f(A) = 1$ if $A \in \mathcal{A}$. Similarly, let g be the characteristic function of $\mathcal{B}$. By the assumptions, f and g are both increasing. Applying the FKG inequality with the trivial measure $\mu \equiv 1$, we get:

$$\Pr(\mathcal{A}\cap\mathcal{B}) = \langle fg\rangle \geq \langle f\rangle\cdot\langle g\rangle = \Pr(\mathcal{A})\cdot\Pr(\mathcal{B}).$$

The other two inequalities follow similarly from Theorem 2.1 and the paragraph following it.

It is worth noting that the proposition can be also derived easily from the Ahlswede–Daykin theorem or from Corollary 1.3. ■

The last proposition has several interesting combinatorial consequences, some of which appeared in Kleitman's (1966b) original paper. Since those are direct combinatorial consequences, and do not contain any additional probabilistic ideas, we omit their exact statement and turn to a version of Proposition 3.1 in a more general probability space.

For a real vector $p = (p_1,\ldots,p_n)$, where $0 \le p_i \le 1$, consider the probability space whose elements are all members of the power set $P(N)$, where, for each $A \subseteq N$,

$$\Pr(A) = \prod_{i\in A} p_i \prod_{j\notin A}(1-p_j).$$

Clearly, this probability distribution is obtained if we choose a random $A \subseteq N$ by choosing each element $i \in N$, independently, with probability p_i. Let us denote, for each $\mathcal{A} \subseteq P(N)$, its probability in this space by $\Pr_p(\mathcal{A})$. In particular, if all the probabilities p_i are 1/2, then $\Pr_p(\mathcal{A})$ is the quantity denoted as $\Pr(\mathcal{A})$ in Proposition 3.1. Define $\mu = \mu_p : P(N) \to \mathbf{R}^+$ by

$$\mu(A) = \prod_{i\in A} p_i \prod_{j\notin A}(1-p_j).$$

It is easy to check that μ is log-supermodular. This is because for $A, B \subseteq N$, $\mu(A)\mu(B) = \mu(A\cup B)\mu(A\cap B)$, as can be checked by comparing the contribution arising from each $i \in N$ to the left-hand side and to the right-hand side of the last equality. Hence one can apply the FKG inequality and obtain the following generalization of Proposition 3.1.

Theorem 3.2. *Let $\mathcal{A}$ and $\mathcal{B}$ be two monotone increasing families of subsets of N and let $\mathcal{C}$ and $\mathcal{D}$ be two monotone decreasing families of subsets of N. Then, for any real vector $p = (p_1,\ldots,p_n)$, $0 \le p_i \le 1$,*

$$\Pr_p(\mathcal{A}\cap\mathcal{B}) \ge \Pr_p(\mathcal{A})\cdot\Pr_p(\mathcal{B}),$$

$$\Pr_p(\mathcal{C}\cap\mathcal{D}) \ge \Pr_p(\mathcal{C})\cdot\Pr_p(\mathcal{D}),$$

$$\Pr_p(\mathcal{A}\cap\mathcal{C}) \le \Pr_p(\mathcal{A})\cdot\Pr_p(\mathcal{C}).$$

This theorem can be applied in many cases and will be used in Chapter 8 to derive the Janson inequalities. As a simple illustration, suppose that $A_1, A_2, \ldots, A_k$ are arbitrary subsets of N and one chooses a random subset A of N by choosing each $i \in N$, independently, with probability p. Then Theorem 3.2 easily implies that

$$\Pr(A \text{ intersects each } A_i) \ge \prod_{i=1}^{k} \Pr(A \text{ intersects } A_i).$$

Notice that this is false, in general, for other similar probabilistic models. For example, if A is a randomly chosen ℓ-element subset of N, then the last inequality may fail.

By viewing the members of N as the $n = \binom{m}{2}$ edges of the complete graph on the set of vertices $V = \{1,2,\ldots,m\}$, we can derive a correlation inequality for random graphs. Let $G = (V,E)$ be a random graph on the set of vertices V generated by choosing, for each $i,j \in V$, $i \neq j$, independently, the pair $\{i,j\}$ to be an edge with probability p. (This model of random graphs is discussed in detail in Chapter 10.) A *property of graphs* is a subset of the set of all graphs on V, closed under isomorphism. Thus, for example, connectivity is a property (corresponding to all connected graphs on V) and planarity is another property. A property Q is *monotone increasing* if whenever G has Q and H is obtained from G by adding edges, then H has Q too. A *monotone decreasing* property is defined in a similar manner. By interpreting the members of N in Theorem 3.2 as the $\binom{m}{2}$ pairs $\{i,j\}$ with $i,j \in V$, $i \neq j$, we obtain:

Theorem 3.3. *Let Q_1, Q_2, Q_3, and Q_4 be graph properties, where Q_1, Q_2 are monotone increasing and Q_3, Q_4 are monotone decreasing. Let $G = (V,E)$ be a random graph on V obtained by picking every edge, independently, with probability p. Then*

$$\Pr(G \in Q_1 \cap Q_2) \geq \Pr(G \in Q_1) \cdot \Pr(G \in Q_2),$$

$$\Pr(G \in Q_3 \cap Q_4) \geq \Pr(G \in Q_3) \cdot \Pr(G \in Q_4),$$

$$\Pr(G \in Q_1 \cap Q_3) \leq \Pr(G \in Q_1) \cdot \Pr(G \in Q_3).$$

Thus, for example, the probability that G is both Hamiltonian and planar does not exceed the product of the probability that it is Hamiltonian by that that it is planar. It seems hopeless to try and prove such a statement directly, without using one of the correlation inequalities.

4. LINEAR EXTENSIONS OF PARTIALLY ORDERED SETS

Let $(P, \leq)$ be a partially ordered set with n elements. A *linear extension* of P is a one-to-one mapping $\sigma : P \to \{1,2,\ldots,n\}$, which is order preserving, i.e., if $x,y \in P$ and $x \leq y$, then $\sigma(x) \leq \sigma(y)$. Intuitively, σ is a ranking of the elements of P which preserves the partial order of P. Consider the probability space of all linear extensions of P, where each possible extension is equally likely. In this space, we can consider events of the form, e.g., $x \leq y$ or $(x \leq y) \wedge (x \leq z)$ (for $x,y,z \in P$), and compute their probabilities. It turns out that the FKG inequality is a very useful tool for studying the correlation between such events. The best known result of this form was conjectured by Rival and Sands and proved by Shepp (1982). It asserts that for any partially ordered set P and any three elements $x,y,z \in P$: $\Pr(x \leq y \wedge x \leq z) \geq \Pr(x \leq y)\Pr(x \leq z)$.

This result became known as the XYZ theorem. Although it looks intuitively obvious, its proof is nontrivial and contains a clever application of the FKG inequality. In this section, we present this result and its elegant proof.

Theorem 4.1. *Let P be a partially ordered set with n elements $a_1, a_2, \ldots, a_n$. Then*

$$\Pr(a_1 \le a_2 \wedge a_1 \le a_3) \ge \Pr(a_1 \le a_2)\Pr(a_1 \le a_3).$$

Proof. Let m be a large integer (which will later tend to infinity) and let L be the set of all ordered n-tuples $\mathbf{x} = (x_1, \ldots, x_n)$, where $x_i \in M = \{1, 2, \ldots, m\}$. (Note that we do *not* assume that the numbers x_i are distinct.) Define an order relation $\le$ on L as follows. For $\mathbf{y} = (y_1, \ldots, y_n) \in L$ and $\mathbf{x}$ as above, $\mathbf{x} \le \mathbf{y}$ iff $x_1 \ge y_1$ and $x_i - x_1 \le y_i - y_1$ for all $2 \le i \le n$. It is not too difficult to check that $(L, \le)$ is a lattice in which the ith component of the meet $\mathbf{x} \wedge \mathbf{y}$ is $(\mathbf{x} \wedge \mathbf{y})_i = \min(x_i - x_1, y_i - y_1) + \max(x_1, y_1)$ and the ith component of the join $\mathbf{x} \vee \mathbf{y}$ is $(\mathbf{x} \vee \mathbf{y})_i = \max(x_i - x_1, y_i - y_1) + \min(x_1, y_1)$.

Moreover, the lattice L is distributive. This follows by an easy computation from the fact that the trivial lattice of integers (with respect to the usual order) is distributive and hence for any three integers a, b, and c,

$$\min(a, \max(b, c)) = \max(\min(a, b), \min(a, c)) \tag{7}$$

and

$$\max(a, \min(b, c)) = \min(\max(a, b), \max(a, c)). \tag{8}$$

Let us show how this implies that L is distributive. Let $\mathbf{x} = (x_1, \ldots, x_n)$, $\mathbf{y} = (y_1, \ldots, y_n)$, and $\mathbf{z} = (z_1, \ldots, z_n)$ be three elements of L. We must show that

$$\mathbf{x} \wedge (\mathbf{y} \vee \mathbf{z}) = (\mathbf{x} \wedge \mathbf{y}) \vee (\mathbf{x} \wedge \mathbf{z}).$$

The ith component of $\mathbf{x} \wedge (\mathbf{y} \vee \mathbf{z})$ is

$$\begin{aligned}(\mathbf{x} \wedge (\mathbf{y} \vee \mathbf{z}))_i &= \min(x_i - x_1, (\mathbf{y} \vee \mathbf{z})_i - (\mathbf{y} \vee \mathbf{z})_1) + \max(x_1, (\mathbf{y} \vee \mathbf{z})_1)\\ &= \min(x_i - x_1, \max(y_i - y_1, z_i - z_1)) + \max(x_1, \min(y_1, z_1)).\end{aligned}$$

Similarly, the ith component of $(\mathbf{x} \wedge \mathbf{y}) \vee (\mathbf{x} \wedge \mathbf{z})$ is

$$\begin{aligned}((\mathbf{x} \wedge \mathbf{y}) \vee (\mathbf{x} \wedge \mathbf{z}))_i &= \max\left((\mathbf{x} \wedge \mathbf{y})_i - (\mathbf{x} \wedge \mathbf{y})_1, (\mathbf{x} \wedge \mathbf{z})_i - (\mathbf{x} \wedge \mathbf{z})_1\right)\\ &\quad + \min\left((\mathbf{x} \wedge \mathbf{y})_1, (\mathbf{x} \wedge \mathbf{z})_1\right)\\ &= \max\left(\min(x_i - x_1, y_i - y_1), \min(x_i - x_1, z_i - z_1)\right)\\ &\quad + \min\left(\max(x_1, y_1), \max(x_1, z_1)\right).\end{aligned}$$

These two quantities are equal, as follows, by applying (7) with $a = x_i - x_1$, $b = y_i - y_1$, $c = z_i - z_1$, and (8) with $a = x_1$, $b = y_1$, $c = z_1$.

Thus L is distributive. To apply the FKG inequality, we need the measure function μ and the two functions f and g. Let μ be the characteristic function of P, i.e., for $\mathbf{x} = (x_1, \ldots, x_n) \in L$, $\mu(\mathbf{x}) = 1$ if $x_i \le x_j$ whenever $a_i \le a_j$ in P, and $\mu(\mathbf{x}) = 0$ otherwise. To show that μ is log-supermodular, it suffices to check that if $\mu(\mathbf{x}) = \mu(\mathbf{y}) = 1$, then $\mu(\mathbf{x} \vee \mathbf{y}) = \mu(\mathbf{x} \wedge \mathbf{y}) = 1$. However, if $\mu(\mathbf{x}) = \mu(\mathbf{y}) = 1$ and $a_i \le a_j$ in P, then $x_i \le x_j$ and $y_i \le y_j$ and therefore

$$\begin{aligned}(x \vee y)_i &= \max(x_i - x_1, y_i - y_1) + \min(x_1, y_1) \\ &\le \max(x_j - x_1,\ y_j - y_1) + \min(x_1, y_1) = (x \vee y)_j,\end{aligned}$$

i.e., $\mu(\mathbf{x} \vee \mathbf{y}) = 1$. Similarly, $\mu(\mathbf{x}) = \mu(\mathbf{y}) = 1$ implies $\mu(\mathbf{x} \wedge \mathbf{y}) = 1$, too.

Not surprisingly, we define the functions f and g as the characteristic functions of the two events $x_1 \le x_2$ and $x_1 \le x_3$, respectively, i.e., $f(\mathbf{x}) = 1$ if $x_1 \le x_2$ and $f(\mathbf{x}) = 0$ otherwise, and $g(\mathbf{x}) = 1$ if $x_1 \le x_3$ and $g(\mathbf{x}) = 0$ otherwise. Trivially, both f and g are increasing. Indeed, if $\mathbf{x} \le \mathbf{y}$ and $f(\mathbf{x}) = 1$, then $0 \le x_2 - x_1 \le y_2 - y_1$ and hence $f(\mathbf{y}) = 1$, and similarly for g.

We therefore have all the necessary ingredients for applying the FKG inequality (Theorem 2.1). This gives the following. Let $(x_1,\ldots,x_n)$ be an n-tuple in L that satisfies the inequalities in P. Then the probability that it satisfies both $x_1 \le x_2$ and $x_1 \le x_3$ is at least as big as the product of the probability that it satisfies $x_1 \le x_2$ by the probability that it satisfies $x_1 \le x_3$. Notice that this is not yet what we wanted to prove; the n-tuples in L are not n-tuples of distinct integers and thus do not correspond to linear extensions of P. However, as $m \to \infty$, the probability that $x_i = x_j$ for some $i \ne j$ in a member $\mathbf{x} = (x_1,\ldots,x_n)$ of L tends to 0 and the assertion of the theorem follows. ■

The Probabilistic Lens: Turán's Theorem

In a graph $G = (V,E)$, let d_v denote the degree of a vertex v and $\alpha(G)$ the maximal size of an independent set of vertices.

Theorem 1.

$$\alpha(G) \geq \sum_{v \in V} \frac{1}{d_v + 1}.$$

Proof. Let $<$ be a uniformly chosen total ordering of V. Define

$$I = \{v \in V : \{v,w\} \in E \Rightarrow v < w\}.$$

Let X_v be the indicator random variable for $v \in I$ and $X = \sum_{v \in V} X_v = |I|$. For each v,

$$E[X_v] = \Pr[v \in I] = \frac{1}{d_v + 1},$$

since $v \in I$ if and only if v is the least element among v and its neighbors. Hence

$$E[X] = \sum_{v \in V} \frac{1}{d_v + 1}$$

and so there exists a specific ordering $<$ with

$$|I| \geq \sum_{v \in V} \frac{1}{d_v + 1}.$$

But if $x,y \in I$ and $\{x,y\} \in E$, then $x < y$ *and* $y < x$, a contradiction. Thus I is independent and $\alpha(G) \geq |I|$. ■

For any $m \leq n$, let q,r satisfy $n = mq + r$, $0 \leq r < m$, and let $e = r\binom{q+1}{2} + (m-r)\binom{q}{2}$. Define a graph $G = G_{n,e}$ on n vertices and e edges by splitting the vertex set into m classes as evenly as possible and joining two vertices if and only if they lie in the same class. Clearly, $\alpha(G_{n,e}) = m$.

Theorem 2 (Turán [1941]). *Let H have n vertices and e edges. Then $\alpha(H) \geq m$ and $\alpha(H) = m \Leftrightarrow H \cong G_{n,e}$.*

Proof. $G_{n,e}$ has $\sum_{v \in V}(d_v + 1)^{-1} = m$, since each clique contributes 1 to the sum. Fixing $e = \sum_{v \in V} d_v/2$, $\sum_{v \in V}(d_v + 1)^{-1}$ is minimized with the d_v as close together as possible. Thus for any H,

$$\alpha(H) \geq \sum_{v \in V} \frac{1}{d_v + 1} \geq m.$$

For $\alpha(H) = m$, we must have equality on both sides above. The second equality implies the d_v must be as close together as possible. Letting $X = |I|$ as in the previous theorem, assume $\alpha(H) = E[X]$. But $\alpha(H) \geq X$ for all values of $<$, so X must be a constant. Suppose H is not a union of cliques. Then there exist $x, y, z \in V$ with $\{x,y\}, \{x,z\} \in E$, $\{y,z\} \notin E$. Let $<$ be an ordering that begins x, y, z and $<'$ the same ordering except that it begins y, z, x, and let I, I' be the corresponding sets of vertices all of whose neighbors are "greater" than it. Then I, I' are identical except that $x \in I$, $y, z \notin I$, whereas $x \notin I'$, $y, z \in I'$. Thus X is not constant. That is, $\alpha(H) = E[X]$ implies that H is the union of cliques and so $H \cong G_{n,e}$. ■

7

Martingales

1. DEFINITIONS

A martingale is a sequence $X_0,\ldots,X_m$ of random variables so that for $0 \le i < m$,

$$E[X_{i+1} \mid X_i] = X_i.$$

Imagine a gambler walking into a casino with X_0 dollars. The casino contains a variety of games of chance. All games are "fair" in that their expectations are zero. The gambler may allow previous history to determine his choice of game and bet. He might employ the gambler's definition of martingale—double the bet until you win. He might play roulette until he wins three times and then switch to keno. Let X_i be the gambler's fortune at time i. Given that $X_i = a$, the conditional expectation of X_{i+1} must be a and so this is a martingale.

A simple but instructive martingale occurs when the gambler plays "flip a coin" for stakes of one dollar each time. Let $Y_1,\ldots,Y_m$ be independent coin flips, each $+1$ or -1 with probability 1/2. Normalize so that $X_0 = 0$ is the gambler's initial stake, though he has unlimited credit. Then $X_i = Y_1 + \cdots + Y_i$ has distribution S_i.

Our martingales will look quite different, at least from the outside.

The Edge Exposure Martingale

Let the random graph $G(n,p)$ be the underlying probability space. Label the potential edges $\{i,j\} \subseteq [n]$ by $e_1,\ldots,e_m$, setting $m = \binom{n}{2}$ for convenience, in any specific manner. Let f be any graph theoretic function. We define a martingale $X_0,\ldots,X_m$ by giving the values $X_i(H)$. $X_m(H)$ is simply $f(H)$. $X_0(H)$ is the expected value of $f(G)$ with $G \sim G(n,p)$. Note that X_0 is a constant. In general (including the cases $i = 0$ and $i = m$),

$$X_i(H) = E[f(G) \mid e_j \in G \longleftrightarrow e_j \in H,\ 1 \le j \le i].$$

In words, to find $X_i(H)$, we first expose the first i pairs $e_1,\ldots,e_i$ and see if they are in H. The remaining edges are not seen and considered to be random. $X_i(H)$ is then the conditional expectation of $f(G)$ with this partial information. When $i = 0$, nothing is exposed and X_0 is a constant. When $i =$

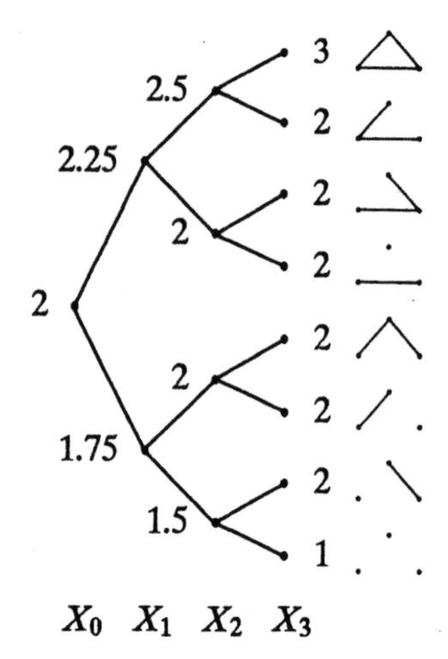

Figure 1. The edge exposure martingale with $n = m = 3, f$ the chromatic number, and the edges exposed in the order "bottom, left, right." The values $X_i(H)$ are given by tracing from the central node to the leaf labeled H.

m, all is exposed and X_m is the function f. The martingale moves from no information to full information in small steps.

Figure 1 shows why this is a martingale. The conditional expectation of $f(H)$ knowing the first $i-1$ edges is the weighted average of the conditional expectations of $f(H)$ where the ith edge has been exposed. More generally—in what is sometimes referred to as a Doob martingale process—X_i may be the conditional expectation of $f(H)$ after certain information is revealed as long as the information known at time i includes the information known at time $i-1$.

The Vertex Exposure Martingale

Again, let $G(n,p)$ be the underlying probability space and f any graph theoretic function. Define $X_1, \ldots, X_n$ by

$$X_i(H) = E[f(G) \mid \text{for } x,y \leq i,\ \{x,y\} \in G \longleftrightarrow \{x,y\} \in H].$$

In words, to find $X_i(H)$, we expose the first i vertices and all their internal edges and take the conditional expectation of $f(G)$ with that partial information. By ordering the edges appropriately, the vertex exposure martingale may be considered a subsequence of the edge exposure martingale. Note that $X_1(H) = E[f(G)]$ is constant, as no edges have been exposed, and $X_n(H) = f(H)$, as all edges have been exposed.

2. LARGE DEVIATIONS

Maurey (1979) applied a large deviation inequality for martingales to prove an isoperimetric inequality for the symmetric group S_n. This inequality was useful in the study of normed spaces; see Milman and Schechtman (1986) for many related results. The applications of martingales in graph theory also all involve the same underlying martingale result used by Maurey, which is the following.

Theorem 2.1 (Azuma's Inequality). *Let $0 = X_0, \ldots, X_m$ be a martingale with*

$$|X_{i+1} - X_i| \leq 1$$

for all $0 \leq i < m$. Let $\lambda > 0$ be arbitrary. Then

$$\Pr\left[X_m > \lambda\sqrt{m}\right] < e^{-\lambda^2/2}.$$

Proof. In the "flip-a-coin" martingale, X_m has distribution S_m and this result is Theorem A.1 in Appendix A. Indeed, the general proof is quite similar. Set, with foresight, $\alpha = \lambda/\sqrt{m}$. Set $Y_i = X_{i+1} - X_i$ so that $|Y_i| \leq 1$ and $E[Y_i|X_{i-1}] = 0$. Then, as in Theorem A.16,

$$E[e^{\alpha Y_i}|X_{i-1}] \leq \cosh(\alpha) \leq e^{\alpha^2/2}.$$

Hence

$$\begin{aligned} E[e^{\alpha X_m}] &= E\left[\prod_{i=1}^{m} e^{\alpha Y_i}\right] \\ &= E\left[\left(\prod_{i=1}^{m-1} e^{\alpha Y_i}\right) E\left(e^{\alpha Y_m} \mid X_{m-1}\right)\right] \\ &\leq E\left[\prod_{i=1}^{m-1} e^{\alpha Y_i}\right] e^{\alpha^2/2} \leq e^{\alpha^2 m/2}, \end{aligned}$$

and

$$\begin{aligned} \Pr\left[X_m > \lambda\sqrt{m}\right] &= \Pr\left[e^{\alpha X_m} > e^{\alpha\lambda\sqrt{m}}\right] \\ &\leq E\left[e^{\alpha X_m}\right] e^{-\alpha\lambda\sqrt{m}} \\ &= e^{\alpha^2 m/2 - \alpha\lambda\sqrt{m}} \\ &= e^{-\lambda^2/2}, \end{aligned}$$

as before. ∎

Corollary 2.2. *Let $c = X_0, \ldots, X_m$ be a martingale with*

$$|X_{i+1} - X_i| \leq 1$$

for all $0 \leq i < m$. Then

$$\Pr\left[|X_m - c| > \lambda\sqrt{m}\right] < 2e^{-\lambda^2/2}.$$

A graph theoretic function f is said to satisfy the edge Lipschitz condition if whenever H and H' differ in only one edge, then $|f(H)-f(H')| \leq 1$. It satisfies the vertex Lipschitz condition if whenever H and H' differ at only one vertex, $|f(H)-f(H')| \leq 1$.

Theorem 2.3. *When f satisfies the edge Lipschitz condition, the corresponding edge exposure martingale satisfies $|X_{i+1}-X_i| \leq 1$. When f satisfies the vertex Lipschitz condition, the corresponding vertex exposure martingale satisfies $|X_{i+1}-X_i| \leq 1$.*

We prove these results in a more general context later. They have the intuitive sense that if knowledge of a particular vertex or edge cannot change f by more than 1, then exposing a vertex or edge should not change the expectation of f by more than 1. Now we give a simple application of these results.

Theorem 2.4 (Shamir and Spencer [1987]). *Let n, p be arbitrary and let $c = E[\chi(G)]$ where $G \sim G(n,p)$. Then*

$$\Pr\left[|\chi(G)-c| > \lambda\sqrt{n-1}\right] < 2e^{-\lambda^2/2}.$$

Proof. Consider the vertex exposure martingale $X_1, \ldots, X_n$ on $G(n,p)$ with $f(G) = \chi(G)$. A single vertex can always be given a new color, so the vertex Lipschitz condition applies. Now apply Azuma's inequality in the form of Corollary 2.2. ■

Letting $\lambda \to \infty$ arbitrarily slowly, this result shows that the distribution of $\chi(G)$ is "tightly concentrated" around its mean. The proof gives no clue as to where the mean is.

3. CHROMATIC NUMBER

In Theorem 3.1 of Chapter 10 we prove that $\chi(G) \sim n/2\log_2 n$ almost surely, where $G \sim G(n, 1/2)$. Here we give the original proof of Béla Bollobás (1988) using martingales. We follow the notations of Chapter 10, Section 3, setting $f(k) = \binom{n}{k}2^{-\binom{k}{2}}$, k_0 so that $f(k_0-1) > 1 > f(k_0)$, $k = k_0 - 4$ so that $k \sim 2\log_2 n$, and $f(k) > n^{3+o(1)}$. Our goal is to show

$$\Pr[\omega(G) < k] = e^{-n^{2+o(1)}},$$

where $\omega(G)$ is the size of the maximum clique of G. We shall actually show in Theorem 3.2 a more precise bound. The remainder of the argument is given in Chapter 10, Section 3.

Let $Y = Y(H)$ be the maximal size of a family of edge disjoint cliques of size k in H. This ingenious and unusual choice of function is key to the martingale proof.

Lemma 3.1.

$$E[Y] \geq \frac{n^2}{2k^4}(1+o(1)).$$

Proof. Let $\mathcal{K}$ denote the family of k-cliques of G so that $f(k)=\mu=E[|\mathcal{K}|]$. Let W denote the number of unordered pairs $\{A,B\}$ of k-cliques of G with $2 \leq |A \cap B| < k$. Then $E[W]=\Delta/2$, with Δ as described in Chapter 10, Section 3 (see also Chapter 4, Section 5), $\Delta \sim \mu^2 k^4 n^{-2}$. Let $\mathcal{C}$ be a random subfamily of $\mathcal{K}$ defined by setting, for each $A \in \mathcal{K}$,

$$\Pr[A \in \mathcal{C}] = q,$$

q to be determined. Let W' be the number of unordered pairs $\{A,B\}$, $A,B \in \mathcal{C}$ with $2 \leq |A \cap B| < k$. Then

$$E[W'] = E[W]q^2 = \frac{\Delta q^2}{2}.$$

Delete from $\mathcal{C}$ one set from each such pair $\{A,B\}$. This yields a set $\mathcal{C}^*$ of edge disjoint k-cliques of G and

$$E[Y] \geq E[|\mathcal{C}^*|] \geq E[|\mathcal{C}|] - E[W'] = \mu q - \frac{\Delta q^2}{2} = \frac{\mu^2}{2\Delta} \sim \frac{n^2}{2k^4},$$

where we choose $q=\mu/\Delta$ (noting that it is less than 1!) to minimize the quadratic. ■

We conjecture that Lemma 3.1 may be improved to $E[Y] > cn^2/k^2$. That is, with positive probability there is a family of k-cliques that are edge disjoint and cover a positive proportion of the edges.

Theorem 3.2.

$$\Pr[\omega(G) < k] < \exp\left(-(c+o(1))\frac{n^2}{\ln^8 n}\right),$$

with c a positive constant.

Proof. Let $Y_0,\ldots,Y_m$, $m=\binom{n}{2}$, be the edge exposure martingale on $G(n, 1/2)$ with the function Y just defined. The function Y satisfies the edge Lipschitz condition, as adding a single edge can only add at most one clique to a family of edge disjoint cliques. (Note that the Lipschitz condition would not be satisfied for the number of k-cliques, as a single edge might yield many new cliques.) G has no k-clique if and only if $Y=0$. Apply Azuma's inequality with $m=\binom{n}{2} \sim n^2/2$ and $E[Y] \geq (n^2/2k^4)(1+o(1))$. Then

$$\begin{aligned}\Pr[\omega(G)<k] = \Pr[Y=0] &\leq \Pr[Y-E[Y] \leq -E[Y]] \leq \exp\left(-E[Y]^2/2\binom{n}{2}\right)\\ &\leq \exp\left(-(c'+o(1))n^2/k^8\right) = \exp\left(-(c+o(1))n^2/\ln^8 n\right)\end{aligned}$$

as desired. ■

Here is another example where the martingale approach requires an inventive choice of graph theoretic function.

Theorem 3.3. *Let $p = n^{-\alpha}$ where α is fixed, $\alpha > 5/6$. Let $G = G(n,p)$. Then there exists $u = u(n,p)$ so that almost always*

$$u \leq \chi(G) \leq u+3.$$

That is, $\chi(G)$ is concentrated in four values.

We first require a well known technical lemma.

Lemma 3.4. *Let α, c be fixed $\alpha > 5/6$. Let $p = n^{-\alpha}$. Then almost always every $c\sqrt{n}$ vertices of $G = G(n,p)$ may be 3-colored.*

Proof of Lemma 3.4. If not, let T be a minimal set that is not 3-colorable. As $T - \{x\}$ is 3-colorable, x must have internal degree of at least 3 in T for all $x \in T$. Thus if T has t vertices, it must have at least $3t/2$ edges. The probability of this occurring for some T with at most $c\sqrt{n}$ vertices is bounded from above by

$$\sum_{t=4}^{c\sqrt{n}} \binom{n}{t} \binom{\binom{t}{2}}{\frac{3t}{2}} p^{3t/2}.$$

We bound

$$\binom{n}{t} \leq \left(\frac{ne}{t}\right)^t \quad \text{and} \quad \binom{\binom{t}{2}}{\frac{3t}{2}} \leq \left(\frac{te}{3}\right)^{3t/2},$$

so each term is at most

$$\left[\frac{ne}{t}\frac{t^{3/2}e^{3/2}}{3^{3/2}}n^{-3\alpha/2}\right]^t \leq \left[c_1 n^{1-3\alpha/2}t^{1/2}\right]^t \leq \left[c_2 n^{1-3\alpha/2}n^{1/4}\right]^t = \left[c_2 n^{-\epsilon}\right]^t$$

with $\epsilon = 3\alpha/2 - 5/4 > 0$ and the sum is therefore $o(1)$. ■

Proof of Theorem 3.3. Let $\epsilon > 0$ be arbitrarily small and let $u = u(n,p,\epsilon)$ be the least integer so that

$$\Pr[\chi(G) \leq u] > \epsilon.$$

Now define $Y(G)$ to be the minimal size of a set of vertices S for which $G - S$ may be u-colored. This Y satisfies the vertex Lipschitz condition, since at worst one could add a vertex to S. Apply the vertex exposure martingale on $G(n,p)$ to Y. Letting $\mu = E[Y]$,

$$\Pr\left[Y \leq \mu - \lambda\sqrt{n-1}\right] < e^{-\lambda^2/2}, \qquad \Pr\left[Y \leq \mu + \lambda\sqrt{n-1}\right] < e^{-\lambda^2/2}.$$

Let λ satisfy $e^{-\lambda^2/2} = \epsilon$, so that these tail events each have probability less than ϵ. We defined u so that with probability at least ϵ, G would be u-colorable and hence $Y = 0$. That is, $\Pr[Y = 0] > \epsilon$. The first inequality therefore forces $\mu \leq \lambda\sqrt{n-1}$. Now, employing the second inequality,

$$\Pr\left[Y \geq 2\lambda\sqrt{n-1}\right] \leq \Pr\left[Y \geq \mu + \lambda\sqrt{n-1}\right] \leq \epsilon.$$

With probability at least $1-\epsilon$, there is a u-coloring of all but at most $c'\sqrt{n}$ vertices. By the lemma almost always, (and certainly with probability at least $1-\epsilon$), these points may be colored with three further colors, giving a $u+3$-coloring of G. The minimality of u guarantees that with probability at least $1-\epsilon$, at least u colors are needed for G. Altogether,

$$\Pr[u \leq \chi(G) \leq u+3] \geq 1-3\epsilon$$

and ϵ was arbitrarily small. ∎

Similar results can be achieved using the same technique for other values of α. For any fixed $\alpha > 1/2$, one finds that $\chi(G)$ is concentrated on some fixed number of values.

4. A GENERAL SETTING

The martingales useful in studying random graphs generally can be placed in the following general setting, which is essentially the one considered in Maurey (1979) and in Milman and Schechtman (1986). Let $\Omega = A^B$ denote the set of functions $g : B \to A$. (With B, the set of pairs of vertices on n vertices and $A = \{0,1\}$ we may identify $g \in A^B$ with a graph on n vertices.) We define a measure by giving values p_{ab} and setting

$$\Pr[g(b) = a] = p_{ab},$$

with the values $g(b)$ assumed mutually independent. (In $G(n,p)$, all $p_{1b} = p$, $p_{0b} = 1-p$.) Now fix a gradation

$$\emptyset = B_0 \subset B_1 \subset \cdots \subset B_m = B.$$

Let $L : A^B \to R$ be a functional (e.g., clique number). We define a martingale $X_0, X_1, \ldots, X_m$ by setting

$$X_i(h) = E[L(g) \mid g(b) = h(b) \text{ for all } b \in B_i].$$

X_0 is a constant, the expected value of L of the random g. X_m is L itself. The values $X_i(g)$ approach $L(g)$ as the values of $g(b)$ are "exposed." We say the functional L satisfies the Lipschitz condition relative to the gradation if for all $0 \leq i < m$,

$$h, h' \text{ differ only on } B_{i+1} - B_i \Rightarrow |L(h') - L(h)| \leq 1.$$

Theorem 4.1. *Let L satisfy the Lipschitz condition. Then the corresponding martingale satisfies*

$$|X_{i+1}(h) - X_i(h)| \leq 1$$

for all $0 \leq i < m$, $h \in A^B$.

Proof. Let H be the family of h' that agrees with h on B_{i+1}. Then

$$X_{i+1}(h) = \sum_{h' \in H} L(h')w_{h'},$$

where $w_{h'}$ is the conditional probability that $g = h'$ given that $g = h$ on B_{i+1}. For each $h' \in H$, let $H[h']$ denote the family of h^* that agrees with h' on all points except (possibly) $B_{i+1} - B_i$. The $H[h']$ partition the family of h^* agreeing with h on B_i. Thus we may express

$$X_i(h) = \sum_{h' \in H} \sum_{h^* \in H[h']} [L(h^*)q_{h^*}]w_{h'},$$

where q_{h^*} is the conditional probability that g agrees with h^* on B_{i+1} given that it agrees with h on B_i. (This is because for $h^* \in H[h']$, $w_{h'}$ is also the conditional probability that $g = h^*$ given that $g = h^*$ on B_{i+1}.) Thus

$$|X_{i+1}(h) - X_i(h)| = \left| \sum_{h' \in H} w_{h'} \left[L(h') - \sum_{h^* \in H[h']} L(h^*)q_{h^*} \right] \right|$$

$$\leq \sum_{h' \in H} w_{h'} \sum_{h^* \in H[h']} \left| q_{h^*} \left[L(h') - L(h^*) \right] \right|.$$

The Lipschitz condition gives $|L(h') - L(h^*)| \leq 1$, so

$$|X_{i+1}(h) - X_i(h)| \leq \sum_{h' \in H} w_{h'} \sum_{h^* \in H[h']} q_{h^*} = \sum_{h' \in H} w_{h'} = 1. \qquad \blacksquare$$

Now we can express Azuma's inequality in a general form.

Theorem 4.2. *Let L satisfy the Lipschitz condition relative to a gradation of length m and let $\mu = E[L(g)]$. Then for all $\lambda > 0$,*

$$\Pr\left[L(g) > \mu + \lambda\sqrt{m}\right] < e^{-\lambda^2/2},$$

$$\Pr\left[L(g) < \mu - \lambda\sqrt{m}\right] < e^{-\lambda^2/2}.$$

5. THREE ILLUSTRATIONS

Let g be the random function from $\{1,\ldots,n\}$ to itself, all n^n possible function equally likely. Let $L(g)$ be the number of values not hit, i.e., the number of y for which $g(x) = y$ has no solution. By linearity of expectation,

$$E[L(g)] = n\left(1 - \frac{1}{n}\right)^n \sim \frac{n}{e}.$$

Set $B_i = \{1,\ldots,i\}$. L satisfies the Lipschitz condition relative to this gradation, since changing the value of $g(i)$ can change $L(g)$ by at most 1. Thus:

Theorem 5.1.

$$\Pr\left[\left|L(g) - \frac{n}{e}\right| > \lambda\sqrt{n}\right] < 2e^{-\lambda^2/2}.$$

Deriving these asymptotic bounds from first principles is quite cumbersome.

As a second illustration, let B be any normed space and let $v_1,\ldots,v_n \in B$ with all $|v_i| \le 1$. Let $\epsilon_1,\ldots,\epsilon_n$ be independent with

$$\Pr[\epsilon_i = +1] = \Pr[\epsilon_i = -1] = \tfrac{1}{2}$$

and set

$$X = |\epsilon_1 v_1 + \cdots + \epsilon_n v_n|.$$

Theorem 5.2.

$$\Pr\left[X - E[X] > \lambda\sqrt{n}\right] < e^{-\lambda^2/2},$$

$$\Pr\left[X - E[X] < -\lambda\sqrt{n}\right] < e^{-\lambda^2/2}.$$

Proof of Theorem 5.2. Consider $\{-1,+1\}^n$ as the underlying probability space, with all $(\epsilon_1,\ldots,\epsilon_n)$ equally likely. Then X is a random variable and we define a martingale $X_0,\ldots,X_n = X$ by exposing one ϵ_i at a time. The value of ϵ_i can only change X by 2, so direct application of Theorem 4.1 gives $|X_{i+1} - X_i| \le 2$. But let ϵ,ϵ' be two n-tuples differing only in the $(i+1)$th coordinate.

$$X_i(\epsilon) = \tfrac{1}{2}\left[X_{i+1}(\epsilon) + X_{i+1}(\epsilon')\right]$$

so that

$$|X_i(\epsilon) - X_{i+1}(\epsilon)| = \tfrac{1}{2}|X_{i+1}(\epsilon') - X_{i+1}(\epsilon)| \le 1.$$

Now apply Azuma's inequality. ■

For a third illustration, let ρ be the Hamming metric on $\{0,1\}^n$. For $A \subseteq \{0,1\}^n$, let $B(A,s)$ denote the set of $y \in \{0,1\}^n$ so that $\rho(x,y) \le s$ for some $x \in A$. ($A \subseteq B(A,s)$ as we may take $x = y$.)

Theorem 5.3. *Let* $\epsilon, \lambda > 0$ *satisfy* $e^{-\lambda^2/2} = \epsilon$. *Then*

$$|A| \geq \epsilon 2^n \Rightarrow \left|B\left(A, 2\lambda\sqrt{n}\right)\right| \geq (1-\epsilon)2^n.$$

Proof. Consider $\{0,1\}^n$ as the underlying probability space, all points equally likely. For $y \in \{0,1\}^n$, set

$$X(y) = \min_{x \in A} \rho(x,y).$$

Let $X_0, X_1, \ldots, X_n = X$ be the martingale given by exposing one coordinate of $\{0,1\}^n$ at a time. The Lipschitz condition holds for X: If y, y' differ in just one coordinate, then $|X(y) - X(y')| \leq 1$. Thus with $\mu = E[X]$,

$$\Pr\left[X < \mu - \lambda\sqrt{n}\right] < e^{-\lambda^2/2} = \epsilon,$$

$$\Pr\left[X > \mu + \lambda\sqrt{n}\right] < e^{-\lambda^2/2} = \epsilon.$$

But

$$\Pr[X = 0] = |A| 2^{-n} \geq \epsilon,$$

so $\mu \leq \lambda\sqrt{n}$. Thus

$$\Pr\left[X > 2\lambda\sqrt{n}\right] < \epsilon$$

and

$$\left|B\left(A, 2\lambda\sqrt{n}\right)\right| = 2^n \Pr\left[X \leq 2\lambda\sqrt{n}\right] \geq 2^n(1-\epsilon).$$ ■

Actually, a much stronger result is known. Let $B(s)$ denote the ball of radius s about $(0, \ldots, 0)$. The isoperimetric inequality proved by Harper in 1966 states that

$$|A| \geq |B(r)| \Rightarrow |B(A,s)| \geq |B(r+s)|.$$

One may actually use this inequality as a beginning to give an alternative proof that $\chi(G) \sim n/2\log_2 n$ and to prove a number of the other results we have shown using martingales.

The Probabilistic Lens: Weierstrass Approximation Theorem

The well-known Weierstrass approximation theorem asserts that the set of real polynomials over $[0,1]$ is dense in the space of all continuous real functions over $[0,1]$. This is stated in the following theorem.

Weierstrass Approximation Theorem. *For every continuous real function* $f : [0,1] \mapsto R$ *and every* $\epsilon > 0$, *there is a polynomial* $p(x)$ *such that* $|p(x) - f(x)| \leq \epsilon$ *for all* $x \in [0,1]$.

Bernstein (1912) gave a charming probabilistic proof of this theorem, based on the properties of the binomial distribution. His proof is the following.

Proof. Since a continuous $f : [0,1] \mapsto R$ is uniformly continuous, there is a $\delta > 0$ such that if $x, x' \in [0,1]$ and $|x - x'| \leq \delta$, then $|f(x) - f(x')| \leq \epsilon/2$. In addition, since f must be bounded, there is an $M > 0$ such that $|f(x)| \leq M$ in $[0,1]$.

Let $B(n,x)$ denote the binomial random variable with n independent trials and probability of success x for each of them. Thus the probability that $B(n,x) = j$ is precisely $\binom{n}{j} x^j (1-x)^{n-j}$. The expectation of $B(n,x)$ is nx and its standard deviation is $\sqrt{nx(1-x)} \leq \sqrt{n}$. Therefore by Chebyschev's inequality, discussed in Chapter 4, for every integer n,

$$\Pr(|B(n,x) - nx| > n^{2/3}) \leq \frac{1}{n^{1/3}}.$$

It follows that there is an integer n such that

$$\Pr\left(|B(n,x) - nx| > n^{2/3}\right) < \frac{\epsilon}{4M}$$

and

$$\frac{1}{n^{1/3}} < \delta.$$

Define

$$P_n(x) = \sum_{i=0}^{n} \binom{n}{i} x^i (1-x)^{n-i} f\left(\frac{i}{n}\right).$$

We claim that for every $x \in [0,1]$, $|P_n(x) - f(x)| \leq \epsilon$. Indeed, since $\sum_{i=0}^{n} \binom{n}{i} x^i (1-x)^{n-i} = 1$, we have

$$\begin{aligned}
|P_n(x) - f(x)| &\leq \sum_{i;|i-nx| \leq n^{2/3}} \binom{n}{i} x^i (1-x)^{n-i} \left| f\left(\frac{i}{n}\right) - f(x) \right| \\
&\quad + \sum_{i;|i-nx| > n^{2/3}} \binom{n}{i} x^i (1-x)^{n-i} \left(\left| f\left(\frac{i}{n}\right) \right| + |f(x)| \right) \\
&\leq \sum_{i;|i/n-x| \leq n^{-1/3} < \delta} \binom{n}{i} x^i (1-x)^{n-i} \left| f\left(\frac{i}{n}\right) - f(x) \right| \\
&\quad + \Pr(|B(n,x) - nx| > n^{2/3}) 2M \\
&\leq \frac{\epsilon}{2} + \frac{\epsilon}{4M} 2M = \epsilon.
\end{aligned}$$

This completes the proof. ■

8

The Poisson Paradigm

When X is the sum of many rare indicator "mostly independent" random variables and $\mu = E[X]$, we would like to say that X is close to a Poisson distribution with mean μ and, in particular, that $\Pr[X = 0]$ is nearly $e^{-\mu}$. We call this rough statement the *Poisson paradigm*. In this chapter, we give a number of situations in which this paradigm may be rigorously proven.

1. THE JANSON INEQUALITIES

In many instances, we would like to bound the probability that none of a set of bad events B_i, $i \in I$, occur. If the events are mutually independent, then

$$\Pr\left[\bigwedge_{i \in I} \overline{B_i}\right] = \prod_{i \in I} \Pr\left[\overline{B_i}\right].$$

When the B_i are "mostly" independent, the Janson inequalities allow us, sometimes, to say that these two quantities are "nearly" equal.

Let Ω be a finite universal set and let R be a random subset of Ω given by

$$\Pr[r \in R] = p_r,$$

these events mutually independent over $r \in \Omega$. Let A_i, $i \in I$, be subsets of Ω, I a finite index set. Let B_i be the *event* $A_i \subseteq R$. (That is, each point $r \in \Omega$ "flips a coin" to determine if it is in R. B_i is the event that the coins for all $r \in A_i$ came up "heads.") Let X_i be the indicator random variable for B_i and $X = \sum_{i \in I} X_i$ the number of $A_i \subseteq R$. The event $\bigwedge_{i \in I} \overline{B_i}$ and $X = 0$ are then identical. For $i, j \in I$, we write $i \sim j$ if $i \neq j$ and $A_i \cap A_j \neq \emptyset$. Note that when $i \neq j$ and not $i \sim j$, then B_i, B_j are independent events, since they involve separate coin flips. Furthermore, and this plays a crucial role in the proofs, if $i \notin J \subset I$ and not $i \sim j$ for all $j \in J$, then B_i is mutually independent of $\{B_j \mid j \in J\}$, i.e., independent of any Boolean function of those B_j. This is because the coin flips on A_i and on $\bigcup_{j \in J} A_j$ are independent. We define

$$\Delta = \sum_{i \sim j} \Pr[B_i \wedge B_j].$$

Here the sum is over ordered pairs, so that $\Delta/2$ gives the same sum over unordered pairs. We set

$$M = \prod_{i\in I} \Pr\left[\overline{B_i}\right],$$

the value of $\Pr[\bigwedge_{i\in I} \overline{B_i}]$ if the B_i were independent.

Theorem 1.1 (The Janson Inequality). *Let B_i, $i \in I$, Δ, M be as above and assume all* $\Pr[B_i] \leq \epsilon$. *Then*

$$M \leq \Pr\left[\bigwedge_{i\in I} \overline{B_i}\right] \leq M \exp\left(\frac{1}{1-\epsilon}\frac{\Delta}{2}\right).$$

Now set

$$\mu = E[X] = \sum_{i\in I} \Pr[B_i].$$

For each $i \in I$,

$$\Pr\left[\overline{B_i}\right] = 1 - \Pr[B_i] \leq \exp(-\Pr[B_i]),$$

so, multiplying over $i \in I$,

$$M \leq e^{-\mu}.$$

It is often more convenient to replace the upper bound of Theorem 1.1 with

$$\Pr\left[\bigwedge_{i\in I} \overline{B_i}\right] \leq \exp\left(-\mu + \frac{1}{1-\epsilon}\frac{\Delta}{2}\right).$$

Perhaps the simplest example of Theorem 1.1 is the asymptotic probability that $G(n, c/n)$ is triangle-free, given in Chapter 10, Section 1. There, as is often the case, $\epsilon = o(1)$, $\Delta = o(1)$, and μ approaches a constant k. In those instances, $\Pr[\bigwedge_{i\in I} \overline{B_i}] \rightarrow e^{-k}$. This is no longer the case when Δ becomes large. Indeed, when $\Delta \geq 2\mu(1-\epsilon)$, the upper bound of Theorem 1.1 becomes useless. Even for Δ slightly less, it is improved by the following result.

Theorem 1.2 (The Generalized Janson Inequality). *Under the assumptions of Theorem* 1.1 *and the further assumption that* $\Delta \geq \mu(1-\epsilon)$,

$$\Pr\left[\bigwedge_{i\in I} \overline{B_i}\right] \leq \exp\left(-\frac{\mu^2(1-\epsilon)}{2\Delta}\right).$$

Theorem 1.2 (when it applies) often gives a much stronger result than Chebyschev's inequality as used in Chapter 4. In Section 3 of that chapter, we saw $\mathrm{var}[X] \leq \mu + \Delta$ so that

$$\Pr\left[\bigwedge_{i\in I} \overline{B_i}\right] = \Pr[X = 0] \leq \frac{\mathrm{var}[X]}{E[X]^2} \leq \frac{\mu+\Delta}{\mu^2}.$$

Suppose $\epsilon = o(1)$, $\mu \to \infty$, $\mu \ll \Delta$, and $\gamma = \mu^2/\Delta \to \infty$. Chebyschev's upper bound on $\Pr[X = 0]$ is then roughly γ^{-1}, while Janson's upper bound is roughly $e^{-\gamma}$.

2. THE PROOFS

The original proofs of Janson are based on estimates of the Laplace transform of an appropriate random variable. The proof we present here follows that of Boppana and Spencer (1989). We shall use the inequalities

$$\Pr\left[B_i \,\middle|\, \bigwedge_{j\in J} \overline{B_j}\right] \le \Pr[B_i]$$

valid for all index sets $J \subset I$, $i \notin J$, and

$$\Pr\left[B_i \mid B_k \wedge \bigwedge_{j\in J} \overline{B_j}\right] \le \Pr[B_i|B_k]$$

valid for all index sets $J \subset I$, $i,k \notin J$. The first follows from Theorem 3.2 of Chapter 6. The second is equivalent to the first, since conditioning on B_k is the same as assuming $p_r = \Pr[r \in R] = 1$ for all $r \in A_k$.

Proof of Theorem 1.1. The lower bound follows immediately. Order the index set $I = \{1,\ldots,m\}$ for convenience. For $1 \le i \le m$,

$$\Pr\left[B_i \,\middle|\, \bigwedge_{1\le j<i} \overline{B_j}\right] \le \Pr[B_i]$$

so

$$\Pr\left[\overline{B_i} \,\middle|\, \bigwedge_{1\le j<i} \overline{B_j}\right] \ge \Pr\left[\overline{B_i}\right]$$

and

$$\Pr\left[\bigwedge_{i\in I} \overline{B_i}\right] = \prod_{i=1}^{m} \Pr\left[\overline{B_i} \,\middle|\, \bigwedge_{1\le j<i} \overline{B_j}\right] \ge \prod_{i=1}^{m} \Pr\left[\overline{B_i}\right].$$

Now the upper bound. For a given i, renumber, for convenience, so that $i \sim j$ for $1 \le j \le d$ and not for $d+1 \le j < i$. We use the inequality $\Pr[A|B \wedge C] \ge \Pr[A \wedge B|C]$, valid for any A,B,C. With $A = B_i$, $B = \overline{B_1} \wedge \cdots \wedge \overline{B_d}$, $C = \overline{B_{d+1}} \wedge \cdots \wedge \overline{B_{i-1}}$,

$$\Pr\left[B_i \,\middle|\, \bigwedge_{1\le j<i} \overline{B_j}\right] = \Pr[A \mid B \wedge C] \ge \Pr[A \wedge B \mid C] = \Pr[A \mid C]\Pr[B \mid A \wedge C].$$

From the mutual independence, $\Pr[A \mid C] = \Pr[A]$. We bound

$$\Pr[B \mid A \wedge C] \geq 1 - \sum_{j=1}^{d} \Pr[B_j \mid B_i \wedge C] \geq 1 - \sum_{j=1}^{d} \Pr[B_j \mid B_i]$$

from the correlation inequality. Thus

$$\Pr\left[B_i \,\middle|\, \bigwedge_{1 \leq j < i} \overline{B_j}\right] \geq \Pr[B_i] - \sum_{j=1}^{d} \Pr[B_j \wedge B_i].$$

Reversing,

$$\Pr\left[\overline{B_i} \,\middle|\, \bigwedge_{1 \leq j < i} \overline{B_j}\right] \leq \Pr\left[\overline{B_i}\right] + \sum_{j=1}^{d} \Pr[B_j \wedge B_i]$$

$$\leq \Pr\left[\overline{B_i}\right] \left(1 + \frac{1}{1-\epsilon} \sum_{j=1}^{d} \Pr[B_j \wedge B_i]\right)$$

since $\Pr[\overline{B_i}] \geq 1 - \epsilon$. Employing the inequality $1 + x \leq e^x$,

$$\Pr\left[\overline{B_i} \,\middle|\, \bigwedge_{1 \leq j < i} \overline{B_j}\right] \leq \Pr\left[\overline{B_i}\right] \exp\left(\frac{1}{1-\epsilon} \sum_{j=1}^{d} \Pr[B_j \wedge B_i]\right)$$

For each $1 \leq i \leq m$, we plug this inequality into

$$\Pr\left[\bigwedge_{i \in I} \overline{B_i}\right] = \prod_{i=1}^{m} \Pr\left[\overline{B_i} \,\middle|\, \bigwedge_{1 \leq j < i} \overline{B_j}\right].$$

The terms $\Pr[\overline{B_i}]$ multiply to M. The exponents add: for each $i, j \in I$ with $j < i$ and $j \sim i$, the term $\Pr[B_j \wedge B_i]$ appears once, so they add to $\Delta/2$. ■

Proof of Theorem 1.2. As discussed in Section 1, the proof of Theorem 1.1 gives

$$\Pr\left[\bigwedge_{i \in I} \overline{B_i}\right] \leq \exp\left(-\mu + \frac{1}{1-\epsilon} \frac{\Delta}{2}\right),$$

which we rewrite as

$$-\ln\left[\Pr\left[\bigwedge_{i \in I} \overline{B_i}\right]\right] \geq \sum_{i \in I} \Pr[B_i] - \frac{1}{2(1-\epsilon)} \sum_{i \sim j} \Pr[B_i \wedge B_j].$$

For any set of indices $S \subset I$, the same inequality applied only to the B_i, $i \in S$, gives

$$-\ln\left[\Pr\left[\bigwedge_{i\in S}\overline{B_i}\right]\right] \geq \sum_{i\in S}\Pr[B_i] - \frac{1}{2(1-\epsilon)}\sum_{i,j\in S, i\sim j}\Pr[B_i \wedge B_j].$$

Let now S be a random subset of I given by

$$\Pr[i \in S] = p,$$

with p a constant to be determined, the events mutually independent. (Here we are using probabilistic methods to prove a probability theorem!) Each term $\Pr[B_i]$ then appears with probability p and each term $\Pr[B_i \wedge B_j]$ with probability p^2 so that

$$E\left[-\ln\left[\Pr\left[\bigwedge_{i\in S}\overline{B_i}\right]\right]\right] \geq E\left[\sum_{i\in S}\Pr[B_i]\right] - \frac{1}{2(1-\epsilon)}E\left[\sum_{i,j\in S, i\sim j}\Pr[B_i \wedge B_j]\right]$$

$$= p\mu - \frac{1}{1-\epsilon}p^2\frac{\Delta}{2}.$$

We set

$$p = \frac{\mu(1-\epsilon)}{\Delta}$$

so as to maximize this quantity. The added assumption of Theorem 1.2 assures us that the probability p is at most 1. Then

$$E\left[-\ln\left[\Pr\left[\bigwedge_{i\in S}\overline{B_i}\right]\right]\right] \geq \frac{\mu^2(1-\epsilon)}{2\Delta}.$$

Therefore there is a specific $S \subset I$ for which

$$-\ln\left[\Pr\left[\bigwedge_{i\in S}\overline{B_i}\right]\right] \geq \frac{\mu^2(1-\epsilon)}{2\Delta}.$$

That is,

$$\Pr\left[\bigwedge_{i\in S}\overline{B_i}\right] \leq \exp\left(-\frac{\mu^2(1-\epsilon)}{2\Delta}\right).$$

But

$$\Pr\left[\bigwedge_{i\in I}\overline{B_i}\right] \leq \Pr\left[\bigwedge_{i\in S}\overline{B_i}\right],$$

completing the proof. ■

3. BRUN'S SIEVE

The more traditional approach to the Poisson paradigm is called *Brun's sieve*, for its use by the number theorist T. Brun. Let $B_1,\ldots,B_m$ be events, X_i the indicator random variable for B_i, and $X = X_1 + \cdots + X_m$ the number of B_i that hold. Let there be a hidden parameter n (so that actually $m = m(n)$, $B_i = B_i^{(n)}$, $X = X^{(n)}$) which will define our o, O notation. Define

$$S^{(r)} = \sum \Pr[B_{i_1} \wedge \cdots \wedge B_{i_r}],$$

the sum over all sets $\{i_1,\ldots,i_r\} \subseteq \{1,\ldots,m\}$. The inclusion–exclusion principle gives that

$$\Pr[X = 0] = \Pr[\overline{B_1} \wedge \cdots \wedge \overline{B_m}] = 1 - S^{(1)} + S^{(2)} - \cdots + (-1)^r S^{(r)} \cdots.$$

Theorem 3.1. *Suppose there is a constant μ so that*

$$E[X] = S^{(1)} \to \mu$$

and such that for every fixed r,

$$E\left[\frac{X^{(r)}}{r!}\right] = S^{(r)} \to \frac{\mu^r}{r!}$$

Then

$$\Pr[X = 0] \to e^{-\mu}$$

and indeed for every t,

$$\Pr[X = t] \to \frac{\mu^t}{t!} e^{-\mu}.$$

Proof. We do only the case $t = 0$. Fix $\epsilon > 0$. Choose s so that

$$\left|\sum_{r=0}^{2s} (-1)^r \frac{\mu^r}{r!} - e^{-\mu}\right| \le \frac{\epsilon}{2}.$$

The Bonferroni inequalities state that, in general, the inclusion–exclusion formula alternately over- and underestimates $\Pr[X = 0]$. In particular,

$$\Pr[X = 0] \le \sum_{r=0}^{2s} (-1)^r S^{(r)}.$$

Select n_0 (the hidden variable) so that for $n \ge n_0$,

$$\left|S^{(r)} - \frac{\mu^r}{r!}\right| \le \frac{\epsilon}{2(2s+1)},$$

for $0 \leq r \leq 2s$. For such n,

$$\Pr[X = 0] \leq e^{-\mu} + \epsilon.$$

Similarly, taking the sum to $2s + 1$, we find n_0 so that for $n \geq n_0$,

$$\Pr[X = 0] \geq e^{-\mu} - \epsilon.$$

As ϵ was arbitrary, $\Pr[X = 0] \to e^{-\mu}$. ■

The threshold functions for $G \sim G(n,p)$ to contain a copy of a given graph H, derived in Chapter 10, Section 1, via the Janson inequality, were originally found using Brun's sieve. Here is an example where both methods are used. Let $G \sim G(n,p)$, the random graph of Chapter 10. Let EPIT represent the statement that *every* vertex lies in a triangle.

Theorem 3.2. *Let $c > 0$ be fixed and let $p = p(n)$, $\mu = \mu(n)$ satisfy*

$$\binom{n-1}{2} p^3 = \mu,$$

$$e^{-\mu} = \frac{c}{n}.$$

Then

$$\lim_{n\to\infty} \Pr[G(n,p) \models \text{EPIT}] = e^{-c}.$$

In Spencer (1990a), threshold functions are found for a very wide class of "extension statements" that assert that every r vertices lie in a copy of some fixed H.

Proof. First fix $x \in V(G)$. For each unordered $y,z \in V(G) - \{x\}$, let B_{xyz} be the event that $\{x,y,z\}$ is a triangle of G. Let C_x be the event $\wedge \overline{B_{xyz}}$ and X_x the corresponding indicator random variable. We use Janson's inequality to bound $E[X_x] = \Pr[C_x]$. Here $p = o(1)$ so $\epsilon = o(1)$. $\sum \Pr[B_{xyz}] = \mu$, as defined above. Dependency $xyz \sim xuv$ occurs if and only if the sets overlap (other than in x). Hence

$$\Delta = \sum_{y,z,z'} \Pr[B_{xyz} \wedge B_{xyz'}] = O(n^3)p^5 = o(1),$$

since $p = n^{-2/3+o(1)}$. Thus

$$E[X_x] \sim e^{-\mu} = \frac{c}{n}.$$

Now define

$$X = \sum_{x \in V(G)} X_x,$$

the number of vertices x not lying in a triangle. Then from linearity of expectation,

$$E[X] = \sum_{x \in V(G)} E[X_x] \to c.$$

We need show that the Poisson paradigm applies to X. Fix r. Then

$$E\left[\frac{X^{(r)}}{r!}\right] = S^{(r)} = \sum \Pr[C_{x_1} \wedge \cdots \wedge C_{x_r}],$$

the sum over all sets of vertices $\{x_1, \ldots, x_r\}$. All r-sets look alike, so

$$E\left[\frac{X^{(r)}}{r!}\right] = \binom{n}{r} \Pr[C_{x_1} \wedge \cdots \wedge C_{x_r}] \sim \frac{n^r}{r!} \Pr[C_{x_1} \wedge \cdots \wedge C_{x_r}],$$

where $x_1, \ldots, x_r$ are some particular vertices. But

$$C_{x_1} \wedge \cdots \wedge C_{x_r} = \wedge \overline{B_{x_i yz}},$$

the conjunction over $1 \le i \le r$ and all y, z. We apply Janson's inequality to this conjunction. Again $\epsilon = p^3 = o(1)$. The number of $\{x_i, y, z\}$ is $r\binom{n-1}{2} - O(n)$, the overcount coming from those triangles containing two (or three) of the x_i. (Here it is crucial that r is fixed.) Thus

$$\sum \Pr[B_{x_i yz}] = p^3 \left(r \binom{n-1}{2} - O(n) \right) = r\mu + O(n^{-1+o(1)}).$$

As before, Δ is p^5 times the number of pairs $x_i yz \sim x_j y'z'$. There are $O(rn^3) = O(n^3)$ terms with $i = j$ and $O(r^2 n^2) = O(n^2)$ terms with $i \neq j$, so again $\Delta = o(1)$. Therefore

$$\Pr[C_{x_1} \wedge \cdots \wedge C_{x_r}] \sim e^{-r\mu}$$

and

$$E\left[\frac{X^{(r)}}{r!}\right] \sim \frac{(ne^{-\mu})^r}{r!} = \frac{c^r}{r!}.$$

Hence the conditions of Theorem 3.1 are met for X. ■

4. LARGE DEVIATIONS

We return to the formulation of Section 1. Our object is to derive large deviation results on X similar to those in Appendix A. Given a point in the probability space (i.e., a selection of R), we call an index set $J \subseteq I$ a disjoint family (abbreviated disfam) if

- B_j for every $j \in J$.
- For no $j, j' \in J$ is $j \sim j'$.

If, in addition,

- If $j' \notin J$ and $B_{j'}$ then $j \sim j'$ for some $j \in J$,

then we call J a maximal disjoint family (maxdisfam). We give some general results on the possible sizes of maxdisfams. The connection to X must then be done on an ad hoc basis.

Lemma 4.1. *With the above notation and for any integer s,*

$$\Pr[\text{there exists a disfam } J,\ |J| = s] \leq \frac{\mu^s}{s!}.$$

Proof. Let $\sum^*$ denote the sum over all s-sets $J \subseteq I$ with no $j \sim j'$. Let $\sum^o$ denote the sum over ordered s-tuples $(j_1, \ldots, j_s)$, with $\{j_1, \ldots, j_s\}$ forming such a J. Let $\sum^a$ denote the sum over *all* ordered s-tuples $(j_1, \ldots, j_s)$. Then

$$\begin{aligned}
\Pr[\text{there exists a disfam } J,\ |J| = s] &\leq \sum{}^* \Pr\left[\bigwedge_{j \in J} B_j\right] \\
&= \sum{}^* \prod_{j \in J} \Pr[B_j] \\
&= \frac{1}{s!} \sum^{o} \Pr[B_{j_1}] \ldots \Pr[B_{j_s}] \\
&\leq \frac{1}{s!} \sum^{a} \Pr[B_{j_1}] \ldots \Pr[B_{j_s}] \\
&\leq \frac{1}{s!} \left[\sum_{i \in I} \Pr[B_i]\right]^s \\
&= \frac{\mu^s}{s!}.
\end{aligned}$$

■

Lemma 4.1 gives an effective upper bound when $\mu^s \ll s!$—basically if $s > \mu\alpha$ for $\alpha > e$. For smaller s, we look at the further conditon of J being a *max*disfam. To that end, we let μ_s denote the minimum, over all $j_1, \ldots, j_s \in I$ of $\sum \Pr[B_i]$, the sum taken over all $i \in I$ except those i with $i \sim j_l$ for some $1 \leq l \leq s$. In application, s will be small (otherwise we use Lemma 4.1) and μ_s will be close to μ. For some applications, it is convenient to set

$$\nu = \max_{j \in I} \sum_{i \sim j} \Pr[B_i]$$

and note that $\mu_s \geq \mu - s\nu$.

Lemma 4.2. *With the above notation and for any integer s,*

$$\Pr[\textit{there exists a maxdisfam } J,\ |J| = s] \le \frac{\mu^s}{s!}\exp(-\mu_s)\exp\left(\frac{\Delta}{2(1-\epsilon)}\right)$$
$$\le \frac{\mu^s}{s!}\exp(-\mu)\exp(s\nu)\exp\left(\frac{\Delta}{2(1-\epsilon)}\right).$$

Proof. As in Lemma 4.1, we bound this probability by $\sum^*$ of $J = \{j_1, \ldots, j_s\}$ being a maxdisfam. For this to occur, J must first be a disfam and then $\bigwedge^* \overline{B_i}$, where $\bigwedge^*$ is the conjunction over all $i \in I$ except those with $i \sim j_l$ for some $1 \le l \le s$. We apply Janson's inequality to give an upper bound to $\Pr[\bigwedge^* \overline{B_i}]$. The associated values μ^*, Δ^* satisfy

$$\mu^* \ge \mu_s, \qquad \Delta^* \le \Delta,$$

the latter, since Δ^* has simply fewer addends. Thus

$$\Pr\left[\bigwedge{}^* \overline{B_i}\right] \le \exp(-\mu_s)\exp\left(\frac{\Delta}{2(1-\epsilon)}\right)$$

and

$$\sum{}^* [\Pr[J \text{ maxdisfam}]] \le \exp(-\mu_s)\exp\left(\frac{\Delta}{2(1-\epsilon)}\right)\sum{}^* \Pr\left[\bigwedge_{j\in J} B_j\right]$$
$$\le \exp(-\mu_s)\exp\left(\frac{\Delta}{2(1-\epsilon)}\right)\frac{\mu^s}{s!}. \qquad \blacksquare$$

When $\epsilon, \Delta = o(1)$ and $\nu\mu = o(1)$, or, more generally, $\mu_{3\mu} \sim \mu$, then Lemma 4.2 gives a close approximation to the Poisson distribution, since

$$\Pr[\text{there exists a maxdisfam } J,\ |J| = s] \le (1 + o(1))\frac{\mu^s}{s!}e^{-\mu}$$

for $s \le 3\mu$ and the probability is quite small for larger s by Lemma 4.1.

5. COUNTING EXTENSIONS

We begin with a case that uses the basic large deviation results of Appendix A.

Theorem 5.1. *Set $p = (\ln n/n)\omega(n)$ where $\omega(n) \to \infty$ arbitrarily slowly. Then in $G(n,p)$ almost always,*

$$\deg(x) \sim (n-1)p$$

for all vertices x.

This is actually a large deviation result. It suffices to show the following.

Theorem 5.2. *Set $p = (\ln n/n)\omega(n)$, where $\omega(n) \to \infty$ arbitrarily slowly. Let $x \in G$ be fixed. Fix $\epsilon > 0$. Then*

$$\Pr[|\deg(x) - (n-1)p| > \epsilon(n-1)p] = o(n^{-1}).$$

Proof. As $\deg(x) \sim B(n-1, p)$, that is, it is a binomial random variable with the above parameters, we have from Theorem A.14 that

$$\Pr[|\deg(x) - (n-1)p| > \epsilon(n-1)p]$$
$$< 2\exp(-c_\epsilon(n-1)p) = o(n^{-1}),$$

as c_ϵ is fixed and $(n-1)p \gg \ln n$. ∎

This result illustrates why logarithmic terms appear so often in the study of random graphs. We want *every* x to have a property; hence we try to get the failure probability down to $o(n^{-1})$. When the Poisson paradigm applies, the failure probability is roughly an exponential, and hence we want the exponent to be logarithmic. This often leads to a logarithmic term for the edge probability p.

In Section 3 we found the threshold function for every vertex to lie on a triangle. It basically occurred when the expected number of extensions of a given vertex to a triangle reached $\ln n$. Now set $N(x)$ to be the number of triangles containing x. Set $\mu = \binom{n-1}{2}p^3 = E[N(x)]$.

Theorem 5.3. *Let p be such that $\mu \gg \ln n$. Then almost always*

$$N(x) \sim \mu$$

for all $x \in G(n,p)$.

As above, this is actually a large deviation result. We actually show the following.

Theorem 5.4. *Let p be such that $\mu \gg \ln n$. Let $x \in G$ be fixed. Fix $\epsilon > 0$. Then*

$$\Pr[|N(x) - \mu| > \epsilon\mu] = o(n^{-1}).$$

Proof. We shall prove this under the further assumption $p = n^{-2/3+o(1)}$ (or, equivalently, $\mu = n^{o(1)}$), which could be removed by technical methods. We now have, in the notation of Lemmas 4.1 and 4.2, $\nu\mu, \epsilon, \Delta = o(1)$. Let P

denote the Poisson distribution with mean μ. Then

$$\Pr[\text{there exists a maxdisfam } J,\ |J| \le \mu(1-\epsilon)]$$
$$\le (1+o(1))\Pr[P \le \mu(1-\epsilon)]$$
$$\Pr[\text{there exists a maxdisfam } J,\ \mu(1+\epsilon) \le |J| \le 3\mu]$$
$$\le (1+o(1))\Pr[\mu(1+\epsilon) \le P \le 3\mu]$$
$$\Pr[\text{there exists a maxdisfam } J,\ |J| \ge 3\mu]$$
$$\le \Pr[\text{there exists a disfam } J,\ |J| \ge 3\mu]$$
$$\le \sum_{s=3\mu}^{\infty} \frac{\mu^s}{s!} = O((1-c)^{\mu}),$$

where $c > 0$ is an absolute constant. Since $\mu \gg \ln n$, the third term is $o(n^{-1})$. The first and second terms are $o(1)$ by Theorem A.15. With probability $1 - o(n^{-1})$, every maxdisfam J has size between $(1-\epsilon)\mu$ and $(1+\epsilon)\mu$.

Fix one such J. (There *always* is some maximal disfam—even if no B_i held, we could take $J = \emptyset$.) The elements of J are triples xyz which form triangles; hence $N(x) \ge |J| \ge (1-\epsilon)\mu$. The upper bound is ad hoc. The probability that there exist five triangles of the form $xyz_1, xyz_2, xyz_3, xyz_4, xyz_5$ is at most $n^6 p^{11} = o(n^{-1})$. The probability that there exist triangles xy_iz_i, xy_iz_i', $1 \le i \le 4$, all vertices distinct, is at most $n^{12}p^{20} = o(n^{-1})$. Consider the graph whose vertices are the triangles xyz, with $\sim$ giving the edge relation. There are $N(x)$ vertices; the maxdisfam J are the maximal independent sets. This graph, with probability $1 - o(n^{-1})$, has no vertex of degree 5 (or more) and no 4 (or more) disjoint edges. This implies that for any J, $|J| \ge N(x) - 12$ and

$$N(x) \le (1+\epsilon)\mu + 12 \le (1+\epsilon')\mu. \qquad \blacksquare$$

For any graph H with "roots" $x_1, \ldots, x_r$, we can examine in $G(n,p)$ the number of extensions $N(x_1, \ldots, x_r)$ of a given set of r vertices to a copy of H. In Spencer (1990b), some general results are given that generalize Theorems 5.2 and 5.4. Under fairly wide assumptions, when the expected number μ of extensions satisfies $\mu \gg \ln n$, then almost always all $N(x_1, \ldots, x_r) \sim \mu$.

6. COUNTING REPRESENTATIONS

The results of this section shall use the following very basic and very useful result.

The Borel–Cantelli Lemma. *Let A_n, $n \in N$, be events with*

$$\sum_{n=1}^{\infty} \Pr[A_n] < \infty.$$

Then

$$\Pr\left[\bigwedge_{i=1}^{\infty}\bigvee_{j=i}^{\infty} A_j\right] = 0.$$

That is, almost always A_n is false for all sufficiently large n. In application, we shall aim for $\Pr[A_n] < n^{-c}$ with $c > 1$ in order to apply this lemma.

Again we begin with a case that involves only the large deviation results of Appendix A. For a given set S of natural numbers, let (for every $n \in N$) $f(n) = f_S(n)$ denote the number of representations $n = x + y$, $x, y \in S$, $x \neq y$.

Theorem 6.1 (Erdős [1956]). *There is a set S for which $f(n) = \Theta(\ln n)$. That is, there is a set S and constants c_1, c_2 so that for all sufficiently large n,*

$$c_1 \ln n \leq f(n) \leq c_2 \ln n.$$

Proof. Define S randomly by

$$\Pr[x \in S] = p_x = \min\left[10\sqrt{\frac{\ln x}{x}}, 1\right].$$

Fix n. Now $f(n)$ is a random variable with mean

$$\mu = E[f(n)] = \sum_{x+y=n} p_x p_y.$$

Roughly there are n addends with $p_x p_y > p_n^2 = 100[(\ln n)/n]$. We have $p_x p_y = \Theta[(\ln n)/n]$ except in the regions $x = o(n), y = o(n)$ and care must be taken that those terms don't contribute significantly to μ. Careful asymptotics (and first year calculus!) yield

$$\mu \sim (100 \ln n) \int_0^1 \frac{dx}{\sqrt{x(1-x)}} = 100\pi \ln n.$$

The negligible effect of the $x = o(n), y = o(n)$ terms reflects the finiteness of the indefinite integral at poles $x = 0$ and $x = 1$. The possible representations $x + y = n$ are mutually independent events so that from Theorem A.14,

$$\Pr[|f(n) - \mu| > \epsilon\mu] < 2(1-\delta)^{\mu}$$

for constants ϵ, δ. To be specific, we take $\epsilon = .9$, $\delta = .1$, and

$$\Pr[|f(n) - \mu| > .9\mu] < (2).9^{314 \ln n} < n^{-1.1}$$

for n sufficiently large. Take $c_1 < .1(100\pi)$ and $c_2 > 1.9(100\pi)$.

Let A_n be the event that $c_1 \ln n \leq f(n) \leq c_2 \ln n$ does *not* hold. We have $\Pr[A_n] < n^{-1.1}$ for n sufficiently large. The Borel–Cantelli lemma applies; almost always all A_n fail for n sufficiently large. Therefore there exists a specific point in the probability space, i.e., a specific set S, for which $c_1 \ln n \leq f(n) \leq c_2 \ln n$ for all sufficiently large n. ∎

The development of the infinite probability space used here, and below, has been carefully done in the book *Sequences* by H. Halberstam and K. F. Roth (1983). The use of the infinite probability space leaves a number of questions about the existential nature of the proof that go beyond the algorithmic. For example, does there exist a recursive set S having the property of Theorem 6.1?

Now for a given set S of natural numbers, let $g(n) = g_S(n)$ denote the number of representations $n = x + y + z$, $x,y,z \in S$, all unequal. The following result was actually proven for representations of n as the sum of k terms for any fixed k. For simplicity, we present here only the proof for $k = 3$.

Theorem 6.2 (Erdős and Tetali [1990]). *There is a set S for which $g(n) = \Theta(\ln n)$. That is, there is a set S and constants c_1, c_2 so that for all sufficiently large n,*

$$c_1 \ln n \leq g(n) \leq c_2 \ln n.$$

Proof. Define S randomly by

$$\Pr[x \in S] = p_x = \min\left[10\left(\frac{\ln x}{x^2}\right)^{1/3}, \frac{1}{2}\right].$$

Fix n. Now $g(n)$ is a random variable and

$$\mu = E[g(n)] = \sum_{x+y+z=n} p_x p_y p_z.$$

Careful asymptotics give

$$\mu \sim 10^3 \ln n \int_{x=0}^{1} \int_{y=0}^{1-x} \frac{dx\,dy}{[xy(1-x-y)]^{2/3}} = K \ln n,$$

where K is large. (We may make K arbitrarily large by increasing "10.") We apply Lemma 4.2. Here $\epsilon = 1/8$ as all $p_x \leq 1/2$. Also,

$$\Delta = \sum p_x p_y p_z p_{y'} p_{z'},$$

the sum over all five-tuples with $x + y + z = x + y' + z' = n$. Roughly, there are n^3 terms, each $\sim p_n^5 = n^{-10/3+o(1)}$ so that the sum is $o(1)$. Again care must be taken that those terms with one (or more) small variables don't contribute much to the sum. We bound $s \leq 3\mu = \Theta(\ln n)$ and consider μ_s. This is the minimal possible $\sum p_x p_y p_z$ over all those x,y,z with $x + y + z = n$ that do not intersect a given s representations; let us weaken that and say a given set of $3s$ elements. Again one needs that the weight of $\sum_{x+y+z=n} p_x p_y p_z$ is not on the edges but "spread" in the center and one shows $\mu_s \sim \mu$. Now, as in Section 5, let P denote the Poisson distribution with mean μ. The probability that there exists a maxdisfam J of size less than $\mu(1-\epsilon)$ or between $\mu(1+\epsilon)$ and 3μ is asymptotically the probability that P lies in that range. For moderate ϵ, as K is large, these—as well as the probability of having a disfam of size bigger than 3μ—will be $o(n^{-c})$ with $c > 1$. By the Borel–Cantelli lemma, almost always all sufficiently large n will have all maxdisfam J of size between $c_1 \ln n$ and $c_2 \ln n$. Then $g(n) \geq c_1 \ln n$ immediately.

The upper bound is again ad hoc. With this p, let $f(n)$ be, as before, the number of representations of n as the sum of two elements of S. We use only that $p_x = x^{-2/3+o(1)}$. We calculate

$$E[f(n)] = \sum_{x+y=n} (xy)^{-2/3+o(1)} = n^{-1/3+o(1)},$$

again watching the "pole" at 0. Here the possible representations are mutually independent, so

$$\Pr[f(n) \geq 4] \leq E[f(n)]^4/4! = n^{-4/3+o(1)},$$

and by the Borel–Cantelli lemma almost always $f(n) \leq 3$ for all sufficiently large n. But then almost always, there is a C so that $f(n) \leq C$ for all n. For all sufficiently large n, there is a maxdisfam (with representations as the sum of three terms) of size less than $c_2 \ln n$. Every triple $x, y, z \in S$ with $x + y + z = n$ must contain at least one of these at most $3c_2 \ln n$ points. The number of triples $x, y, z \in S$ with $x + y + z = n$ for a particular x is simply $f(n - x)$, the number of representations $n - x = y + z$ (possibly 1 less, since $y, z \neq x$), and so is at most C. But then there are at most $C(3c_2 \ln n)$ total representations $n = x + y + z$. ■

7. FURTHER INEQUALITIES

Here we discuss some further results that allow one, sometimes, to apply the Poisson paradigm. Let B_i, $i \in I$, be events in an arbitrary probability space. As in the Lovász local lemma of Chapter 5, we say that a binary relation $\sim$ on I is a *dependency digraph* if for each $i \in I$ the event B_i is mutually independent of $\{B_j \mid \text{not } i \sim j\}$. (The digraph of Chapter 5, Section 1, has $E = \{(i,j) \mid i \sim j\}$.) Suppose the events B_i satisfy the inequalities of Section 2:

$$\Pr\left[B_i \,\middle|\, \bigwedge_{j \in J} \overline{B_j}\right] \leq \Pr[B_i]$$

valid for all index sets $J \subset I$, $i \notin J$ and

$$\Pr\left[B_i \mid B_k \wedge \bigwedge_{j \in J} \overline{B_j}\right] \leq \Pr[B_i | B_k]$$

valid for all index sets $J \subset I$, $i,k \notin J$. Then the Janson inequalities, Theorems 1.1 and 1.2, and also Lemmas 4.1 and 4.2 hold as stated. The proofs are identical; the above are the only properties of the events B_i that were used.

Suen (1990) has given a very general result that allows the approximation of $\Pr[\bigwedge_{i \in I} \overline{B_i}]$ by $M = \prod_{i \in I} \Pr[\overline{B_i}]$. Again let B_i, $i \in I$, be events in an arbitrary probability space. We say that a binary relation $\sim$ on I is a *superdependency digraph* if the following holds: Let $J_1, J_2 \subset I$ be disjoint subsets so that $j_1 \sim j_2$ for no $j_1 \in J_1$, $j_2 \in J_2$. Let B^1 be any Boolean combination of the events $B_j, j \in J_1$, and let B^2 be any Boolean combination of the events $B_j, j \in J_2$. Then B^1, B^2 are independent. Note that the $\sim$ of Section 1 is indeed a superdependency digraph.

Theorem 7.1 (Suen [1990]). *Under the above conditions,*

$$\left| \Pr\left[\bigwedge_{i \in I} \overline{B_i} \right] - M \right| \le M \left[\exp\left(\sum_{i \sim j} y(i,j) \right) - 1 \right],$$

where

$$y(i,j) = (\Pr[B_i \wedge B_j] + \Pr[B_i]\Pr[B_j]) \prod_{l \sim i \text{ or } l \sim j} (1 - \Pr[B_l])^{-1}.$$

We shall not prove Theorem 7.1. In many instances, the above product is not large. Suppose it is less than 2 for all $i \sim j$. In that case,

$$\sum_{i \sim j} y(i,j) \le 2 \left[\Delta + \sum_{i \sim j} \Pr[B_i]\Pr[B_j] \right].$$

In many instances $\sum_{i \sim j} \Pr[B_i]\Pr[B_j]$ is small relative to Δ (as in many instances when $i \sim j$, the events B_i, B_j are positively correlated). When, furthermore, $\Delta = o(1)$, Suen's theorem gives the approximation of $\Pr[\bigwedge_{i \in I} \overline{B_i}]$ by M. Suen has applied this result to examinations of the number of *induced* copies of a fixed graph H in the random $G(n,p)$.

Janson (1990) has given a one-way large deviation result on the X of Section 1 which is somewhat simpler to apply than Lemmas 4.1 and 4.2.

Theorem 7.2 (Janson). *With $\mu = E[X]$ and $\gamma > 0$ arbitrary,*

$$\Pr[X \le (1-\gamma)\mu] < \exp\left(\frac{-\gamma^2 \mu}{2 + \frac{\Delta}{\mu}} \right).$$

When $\Delta = o(\mu)$, this bound on the tail approximates that of the normal curve with mean and standard deviation μ. We shall not prove Theorem 7.2 here.

The proofs of Theorems 7.1 and 7.2 as well as the original proofs by Janson of Theorems 1.1 and 1.2 are based on estimations of the Laplace transform of X, bounding $E[e^{-tX}]$.

The Probabilistic Lens: Local Coloring

This result of Erdős (1962) gives further probabilistic evidence that the chromatic number $\chi(G)$ cannot be deduced from local considerations.

Theorem. *For all k, there exists $\epsilon > 0$ so that for all sufficiently large n, there exist graphs G on n vertices with $\chi(G) > k$ and yet $\chi(G|_S) \le 3$ for every set S of vertices of size at most ϵn.*

Proof. For a given k, let $c, \epsilon > 0$ satisfy (with foresight)

$$c > 2k^2 H\left(\frac{1}{k}\right) \ln 2,$$

$$\epsilon < e^{-5} 3^3 c^{-3},$$

where $H(x) = -x \log_2 x - (1-x)\log_2(1-x)$ is the entropy function. Set $p = c/n$ and let $G \sim G(n,p)$. We show that G almost surely satisfies the two conditions of the theorem.

If $\chi(G) \le k$, there would be an independent set of size n/k. The expected number of such sets is

$$\binom{n}{n/k}(1-p)^{\binom{n/k}{2}} < 2^{n(H(1/k)+o(1))} \exp\left(\frac{-cn}{2k^2(1+o(1))}\right)$$

which is $o(1)$ by our condition on c. Hence almost surely, $\chi(G) > k$.

Suppose some set S with $t \le \epsilon n$ vertices required at least four colors. Then there would be a minimal such set S. For any $v \in S$, there would be a 3-coloring of $S - \{v\}$. If v had two or fewer neighbors in S then this could be extended to a 3-coloring of S. Hence every $v \in S$ would have degree at least 3 in $G|_S$ and so $G|_S$ would have at least $3t/2$ edges. The probability that some $t \le \epsilon n$ vertices have at least $3t/2$ edges is less than

$$\sum_{t \le \epsilon n} \binom{n}{t} \binom{\binom{t}{2}}{3t/2} \left(\frac{c}{n}\right)^{3t/2}.$$

We outline the analysis. When $t = O(1)$, the terms are negligible. Otherwise we bound each term from above by

$$\left[\frac{ne}{t}\left(\frac{te}{3}\right)^{3/2}\left(\frac{c}{n}\right)^{3/2}\right]^t \leq \left[e^{5/2}3^{-3/2}c^{3/2}\sqrt{t/n}\right]^t.$$

Now since $t \leq \epsilon n$, the bracketed term is at most $e^{5/2}3^{-3/2}c^{3/2}\epsilon^{1/2}$, which is less than 1 by our condition on ϵ. The full sum is $o(1)$, i.e., almost surely no such S exists. ■

Many tempting conjectures are easily *dis*proved by the probabilistic method. If every $n/(\ln n)$ vertices may be 3-colored, then can a graph G on n vertices be 4-colored? This result shows that the answer is no.

9

Pseudo-Randomness

As shown in the various chapters of this book, the probabilistic method is a powerful tool for establishing the existence of combinatorial structures with certain properties. It is often the case that such an existence proof is not sufficient; we actually prefer an *explicit construction*. This is not only because an explicit construction may shed more light on the corresponding problem, but also because it often happens that a random-looking structure is useful for a certain algorithmic procedure; in this case, we would like to have an algorithm and not merely to prove that it exists.

The problem of finding explicit constructions may look trivial; after all, since we are mainly dealing with finite cases, once we have a probabilistic proof of existence we can find an explicit example by exhaustive search. Moreover, many of the probabilistic proofs of existence actually show that most members of a properly chosen random space have the desired properties. We may thus expect that it would not be too difficult to find one such member. Although this is true in principle, it is certainly not practical to check all possibilities; it is thus common to define an explicit construction of a combinatorial object as one that can be performed efficiently; say, in time that is polynomial in the parameters of the object.

Let us illustrate this notion by one of the best known open problems in the area of explicit constructions; the problem of constructing explicit *Ramsey graphs*. The first example given in Chapter 1 is the proof of Erdős that for every n there are graphs on n vertices containing neither a clique nor an independent set on $2\log_2 n$ vertices. This proof is an existence proof; can we actually describe explicitly such graphs? Erdős offered a prize of $500 for the explicit construction of an infinite family of graphs, in which there is neither a clique nor an independent set of size more than a constant times the logarithm of the number of vertices, for some absolute constant. Of course, we can, in principle, for every fixed n, check all graphs on n vertices until we find a good one, but this does not give an efficient way of producing the desired graphs and hence is not explicit. Although the problem mentioned above received a considerable amount of attention, it is still open. The best known explicit construction is due to Frankl and Wilson (1981), who describe explicit graphs on n vertices that contain neither a clique nor an independent set on more than $2^{c\sqrt{\log n \log\log n}}$ vertices, for some absolute positive constant c.

Although the problem of constructing explicit Ramsey graphs is still open, there are several other problems, for which explicit constructions are known. In this chapter, we present a few examples and discuss briefly some of their algorithmic applications. We also describe several seemingly unrelated properties of a graph, which all turn out to be equivalent. All these are properties of the random graph and it is thus common to call a graph that satisfies these properties *quasi-random*. The equivalence of all these properties enables one to show, in several cases, that certain explicit graphs have many pseudo-random properties by merely showing that they possess one of them. These notions have been generalized to hypergraphs, and a beautiful related result of Rödl (1985) on asymptotic covering is presented in the last section of the chapter.

1. THE QUADRATIC RESIDUE TOURNAMENTS

Recall that a *tournament* on a set V of n players is an orientation $T = (V,E)$ of the set of edges of the complete graph on the set of vertices V. If (x,y) is a directed edge, we say that x *beats* y. Given a permutation π of the set of players, a (directed) edge (x,y) of the tournament is *consistent* with π if x precedes y in π. If π is viewed as a ranking of the players, then it is reasonable to try and find rankings with as many consistent arcs as possible. Let $c(\pi,T)$ denote the number of arcs of T that are consistent with π, and define $c(T) = \max(c(\pi,T))$, where the maximum is taken over all permutations π of the set of vertices of T. For every tournament T on n players, if $\pi = 1,2,\ldots,n$ and $\pi' = n,n-1,\ldots,1$, then $c(\pi,T)+c(\pi',T) = \binom{n}{2}$. Therefore $c(T) \geq \frac{1}{2}\binom{n}{2}$. In fact, it can be shown that for every such T, $c(T) \geq \frac{1}{2}\binom{n}{2} + \Omega(n^{3/2})$. On the other hand, a simple probabilistic argument shows that there are tournaments T on n players for which $c(T) \leq (1+o(1))\frac{1}{2}\binom{n}{2}$. (The best known estimate, which gives the right order of magnitude for the largest possible value of the difference of $c(T) - \frac{1}{2}\binom{n}{2}$, is more complicated and was given by de la Vega in 1983. He showed that there are tournaments T on n players for which $c(T) \leq \frac{1}{2}\binom{n}{2} + O(n^{3/2})$.)

Can we explicitly describe tournaments T on n vertices in which $c(T) \leq (1+o(1))\frac{1}{2}\binom{n}{2}$? This problem was mentioned by Erdős and Moon (1965) and by Spencer (1985). It turns out that several such constructions can be given. Let us describe one.

Let $p \equiv 3 \pmod 4$ be a prime and let $T = T_p$ be the tournament whose vertices are all elements of the finite field $GF(p)$ in which (i,j) is a directed edge iff $i-j$ is a quadratic residue. (Since $p \equiv 3 \pmod 4$, -1 is a quadratic nonresidue modulo p and hence T_p is a well-defined tournament.)

Theorem 1.1. *For the tournaments T_p described above,*

$$c(T_p) \leq \frac{1}{2}\binom{p}{2} + O(p^{3/2}\log p).$$

In order to prove this theorem, we need some preparations. Let χ be the quadratic residue character defined on the elements of the finite field $GF(p)$ by $\chi(y) = y^{(p-1)/2}$. Equivalently, $\chi(y)$ is 1 if y is a nonzero square, 0 if y is 0, and -1 otherwise. Let $D = (d_{ij})_{i,j=0}^{p-1}$ be the p by p matrix defined by $d_{ij} = \chi(i-j)$.

Fact. *For every two distinct j and l, $\sum_{i \in GF(p)} d_{ij} d_{il} = -1$.*

Proof of Fact.

$$\sum_i d_{ij} d_{il} = \sum_i \chi(i-j)\chi(i-l)$$

$$= \sum_{i \neq j,l} \chi(i-j)\chi(i-l)$$

$$= \sum_{i \neq j,l} \chi\left(\frac{i-j}{i-l}\right)$$

$$= \sum_{i \neq j,l} \chi\left(1 + \frac{l-j}{i-l}\right).$$

As i ranges over all elements of $GF(p)$ besides j and l, the quantity $(1 + (l-j)/(i-l))$ ranges over all elements of $GF(p)$ besides 0 and 1. Since the sum of $\chi(r)$ over all r in $GF(p)$ is 0, this implies that the right-hand side of the last equation is $0 - \chi(0) - \chi(1) = -1$, completing the proof. ■

For two subsets A and B of $GF(p)$, let $e(A,B)$ denote the number of directed edges of T_p that start in a vertex of A and end in a vertex of B. By the definition of the matrix D, it follows that

$$\sum_{i \in A} \sum_{j \in B} d_{ij} = e(A,B) - e(B,A).$$

The following lemma is proved in Alon (1986a).

Lemma 1.2. *For any two subsets A and B of $GF(p)$,*

$$\left| \sum_{i \in A} \sum_{j \in B} d_{ij} \right| \leq |A|^{1/2} |B|^{1/2} p^{1/2}.$$

Proof of Lemma 1.2. By the Cauchy–Schwarz inequality and by the fact above,

$$\left(\sum_{i\in A}\sum_{j\in B} d_{ij}\right)^2 \le |A|\left(\sum_{i\in A}\left(\sum_{j\in B} d_{ij}\right)^2\right)$$

$$\le |A|\left(\sum_{i\in GF(p)}\left(\sum_{j\in B} d_{ij}\right)^2\right)$$

$$= |A|\left(\sum_{i\in GF(p)}\left(|B| + 2\sum_{j<l\in B} d_{ij}d_{il}\right)\right)$$

$$= |A|\,|B|p + 2|A|\sum_{j<l\in B}\sum_{i\in GF(P)} d_{ij}d_{il}$$

$$= |A|\,|B|p - |A|\,|B|(|B|-1) = |A|\,|B|(p-|B|+1) \le |A|\,|B|p,$$

completing the proof of the lemma. ■

Proof of Theorem 1.1. Let r be the smallest integer satisfying $2^r \ge p$. Let $\pi = \pi_1,\ldots,\pi_p$ be an arbitrary permutation of the vertices of T_p, and define $\pi' = \pi_p,\ldots,\pi_1$. We must show that $c(\pi,T_p) \le 1/2\binom{p}{2} + O(p^{3/2}\log p)$, or eqivalently, that $c(\pi,T_p) - c(\pi',T_p) \le O(p^{3/2}\log p)$. Let a_1 and a_2 be two integers satisfying $p = a_1 + a_2$ and $a_1 \le 2^{r-1}$, $a_2 \le 2^{r-1}$. Let A_1 be the set of the first a_1 vertices in the permutation π and let A_2 be the set of the last a_2 vertices in π. By Lemma 1.2,

$$e(A_1,A_2) - e(A_2,A_1) \le (a_1a_2p)^{1/2} \le 2^{r-1}p^{1/2}.$$

Next, let $a_{11},a_{12},a_{21},a_{22}$ be integers, each of which does not exceed 2^{r-2} such that $a_1 = a_{11} + a_{12}$ and $a_2 = a_{21} + a_{22}$. Let A_{11} be the subset of A_1 consisting of those a_{11} elements of A_1 that appear first in π, and let A_{12} be the set of the a_{12} remaining elements of A_1. The partition of A_2 into the two sets A_{21} and A_{22} is defined similarly. By applying Lemma 1.2, we obtain

$$e(A_{11},A_{12}) - e(A_{12},A_{11}) + e(A_{21},A_{22}) - e(A_{22},A_{21})$$
$$\le (a_{11}a_{12}p)^{1/2} + (a_{21}a_{22}p)^{1/2} \le 2\cdot 2^{r-2}p^{1/2}.$$

Continuing in the same manner, we obtain, in the ith step, a partition of the set of vertices into 2^i blocks, each consisting of at most 2^{r-i} consecutive elements in the permutation π. This partition is obtained by splitting each block in the partition corresponding to the previous step into two parts. By applying Lemma 1.2 to each such pair $A_{\epsilon 1}$, $A_{\epsilon 2}$ (where here ϵ is a vector of length $i-1$ with $\{1,2\}$-entries), and by summing, we conclude that the sum over all these 2^{i-1} vectors ϵ of the differences $e(A_{\epsilon 1},A_{\epsilon 2})-e(A_{\epsilon 2},A_{\epsilon 1})$ does not exceed

$$2^{i-1}2^{r-i}p^{1/2}\le 2^{r-1}p^{1/2}.$$

Observe that the sum of the left-hand sides of all these inequalities as i ranges from 1 to r is precisely the difference $c(\pi,T_p)-c(\pi',T_p)$. Therefore by summing we obtain

$$c(\pi,T_p)-c(\pi',T_p)\le 2^{r-1}p^{1/2}r = O(p^{3/2}\log p),$$

completing the proof. ■

We note that any antisymmetric matrix with $\{1,-1\}$-entries in which each two rows are roughly orthogonal can be used to give a construction of a tournament as above. Some related results appear in Frankl, Rödl, and Wilson (1988). The tournaments T_p, however, have stronger pseudo-random properties than do some of these other tournaments. For example, for every $k\le 1/4\log p$, and for every set S of k vertices of T_p, the number of vertices of T_p that beat all the members of S is $(1+o(1))p/2^k$. This was proved by Graham and Spencer (1971) by applying Weil's famous theorem, known as the Riemann hypotheses for curves over finite fields (Weil [1948]). Taking a sufficiently large p, this supplies an explicit construction for the Schütte problem mentioned in Chapter 1.

2. EIGENVALUES AND EXPANDERS

A graph $G=(V,E)$ is called an *(n,d,c)-expander* if it has n vertices, the maximum degree of a vertex is d, and for every set of vertices $W\subset V$ of cardinality $|W|\le n/2$, the inequality $|N(W)|\ge c|W|$ holds, where $N(W)$ denotes the set of all vertices in $V\backslash W$ adjacent to some vertex in W. We note that sometimes a slightly different definition is used, but the difference is not essential. Expanders share many of the properties of sparse random graphs, and are the subject of an extensive literature. A family of *linear expanders of density d and expansion c* is a sequence $\{G_i\}_{i=1}^{\infty}$, where G_i is an (n_i,d,c)-expander and n_i tends to infinity as i tends to infinity.

Such a family is the main component of the parallel sorting network of Ajtai, Komlós, and Szemerédi (1983), and can be used for constructing certain fault-tolerant linear arrays. It also forms the basic building block used in the construction of graphs with special connectivity properties and small number of edges. Some other examples of the numerous applications of these graphs to various problems in theoretical computer science can be found in, e.g., Alon (1986a) and its references.

It is not too hard to prove the existence of a family of linear expanders using probabilistic arguments. This was first done by Pinsker (1973). An explicit construction is much more difficult to find, and was first given by Margulis (1973). This construction was later improved by various authors; most known constructions are Cayley graphs of certain groups of matrices, and their expansion properties are proved by estimating the eigenvalues of the adjacency matrices of the graphs and by relying on the close correspondence between the expansion properties of a graph and its spectral properties. This correspondence was first studied, independently, by Tanner (1984) and by Alon and Milman (1984). Since it is somewhat simpler for the case of regular graphs, we restrict our attention here to this case.

Let $G = (V,E)$ be a d-regular graph and let $A = A_G = (a_{uv})_{u,v \in V}$ be its adjacency matrix given by $a_{uv} = 1$ if $uv \in E$ and $a_{uv} = 0$ otherwise. Since G is d-regular, the largest eigenvalue of A is d, corresponding to the all-1 eigenvector. Let $\lambda = \lambda(G)$ denote the second largest eigenvalue of G. For two (not necessarily disjoint) subsets B and C of V, let $e(B,C)$ denote the number of ordered pairs (u,v), where $u \in B$, $v \in C$, and uv is an edge of G. (Note that if B and C are disjoint, this is simply the number of edges of G that connect a vertex of B with a vertex of C.)

Theorem 2.1. *For every partition of the set of vertices V into two disjoint subsets B and C,*

$$e(B,C) \geq \frac{(d-\lambda)|B||C|}{n}.$$

Proof. Put $|V| = n$, $b = |B|$, $c = |C| = n - b$. Let $D = dI$ be the n by n scalar matrix with the degree of regularity of G on its diagonal. Observe that for any real vector x of length n (considered as a function $x : V \mapsto R$), we have

$$\begin{aligned}((D-A)x,x) &= \sum_{u \in V}\Big(d(x(u))^2 - \sum_{v;uv \in E} x(v)x(u)\Big)\\ &= d\sum_{u \in V}(x(u))^2 - 2\sum_{uv \in E} x(v)x(u)\\ &= \sum_{uv \in E}(x(v)-x(u))^2.\end{aligned}$$

Define, now, a vector x by $x(v) = -c$ if $v \in B$ and $x(v) = b$ if $v \in C$. Notice that A and $D - A$ have the same eigenvectors , and that the eigenvalues of $D - A$ are precisely $d - \mu$, as μ ranges over all eigenvalues of A. Note, also, that $\sum_{v \in V} x(v) = 0$, i.e., x is orthogonal to the eigenvector of the smallest eigenvalue of $D - A$. Since $D - A$ is a symmetric matrix, its eigenvectors are orthogonal to each other and form a basis of the n-dimensional space. It follows that x is a linear combination of the other eigenvectors of $D - A$ and hence, by the definition of λ and the fact that $d - \lambda$ is the second smallest eigenvalue of $D - A$, we conclude that $((D-A)x,x) \geq (d-\lambda)(x,x) = (d-\lambda)(bc^2 + cb^2) = (d-\lambda)bcn$.

By the second paragraph of the proof, the left-hand side of the last inequality is $\sum_{uv \in E}(x(u)-x(v))^2 = e(B,C)\cdot(b+c)^2 = e(B,C)\cdot n^2$. Thus

$$e(B,C) \geq \frac{(d-\lambda)bc}{n},$$

completing the proof. ■

Corollary 2.2. *If λ is the second largest eigenvalue of a d-regular graph G with n vertices, then G is an (n,d,c)-expander for $c = (d-\lambda)/2d$.*

Proof. Let W be a set of $w \leq n/2$ vertices of G. By Theorem 2.1, there are at least $[(d-\lambda)w(n-w)]/n \geq [(d-\lambda)w]/2$ edges from W to its complement. Since no vertex in the complement is adjacent to more than d of these edges, it follows that $|N(W)| \geq [(d-\lambda)w]/2d$. ■

The estimate for c in the last corollary can in fact be improved to $[2(d-\lambda)]/(3d-2\lambda)$, as shown by Alon and Milman (1984). Each of these estimates shows that if the second largest eigenvalue of G is far from the first, then G is a good expander. The converse of this is also true, although more complicated. This is given in the following result, proved in Alon (1986b), which we state without its proof.

Theorem 2.3. *If G is a d-regular graph that is an (n,d,c)-expander, then $\lambda(G) \leq d - c^2/(4+2c^2)$.*

The last two results supply an efficient algorithm for approximating the expanding properties of a d-regular graph; we simply compute (or estimate) its second largest eigenvalue. The larger the difference between this eigenvalue and d is, the better the expanding properties of G are. It is thus natural to ask how far from d this second eigenvalue can be. It is known (see Nilli [1990]) that the second largest eigenvalue of any d-regular graph with diameter k is at least $2\sqrt{d-1}(1-O(1/k))$. Therefore, in any infinite family of d-regular graphs, the limsup of the second largest eigenvalue is at least $2\sqrt{d-1}$. Lubotzky, Phillips, and Sarnak (1986), and independently, Margulis (1988), gave, for every $d = p+1$ where p is a prime congruent to 1 modulo 4, explicit constructions of infinite families of d-regular graphs G_i with second largest eigenvalues $\lambda(G_i) \leq 2\sqrt{d-1}$. These graphs are Cayley graphs of factor groups of the group of all 2 by 2 invertible matrices over a finite field, and their eigenvalues are estimated by applying results of Eichler and Igusa concerning the Ramanujan conjecture. Eichler's proof relies on Weil's theorem mentioned in the previous section. The non-bipartite graphs G constructed in this manner satisfy a somewhat stronger assertion than $\lambda(G) \leq 2\sqrt{d-1}$. In fact, besides their largest eigenvalue d, they do not have eigenvalues whose absolute value exceeds $2\sqrt{d-1}$. This fact implies some strong pseudo-random properties, as shown in the next results.

Theorem 2.4. *Let $G = (V, E)$ be a d-regular graph on n vertices, and suppose the absolute value of each of its eigenvalues but the first one is at most λ. For a vertex $v \in V$ and a subset B of V, denote by $N(v)$ the set of all neighbors of v in G, and let $N_B(v) = N(v) \cap B$ denote the set of all neighbors of v in B. Then, for every subset B of cardinality bn of V,*

$$\sum_{v \in V} (|N_B(v)| - bd)^2 \leq \lambda^2 b(1-b)n.$$

Observe that in a random d-regular graph, each vertex v would tend to have about bd neighbors in each set of size bn. The above theorem shows that if λ is much smaller than d, then for most vertices v, $N_B(v)$ is not too far from bd.

Proof. Let A be the adjacency matrix of G and define a vector $f : V \mapsto R$ by $f(v) = 1 - b$ for $v \in B$ and $f(v) = -b$ for $v \notin B$. Clearly, $\sum_{v \in V} f(v) = 0$, i.e., f is orthogonal to the eigenvector of the largest eigenvalue of A. Therefore

$$(Af, Af) \leq \lambda^2 (f, f).$$

The right-hand side of the last inequality is $\lambda^2(bn(1-b)^2 + (1-b)nb^2) = \lambda^2 b(1-b)n$. The left-hand side is

$$\sum_{v \in V} ((1-b)|N_B(v)| - b(d - |N_B(v)|))^2 = \sum_{v \in V} (|N_B(v)| - bd)^2.$$

The desired result follows. ■

Corollary 2.5. *Let $G = (V, E)$, d, n, and λ be as in Theorem 2.4. Then for every two sets of vertices B and C of G, where $|B| = bn$ and $|C| = cn$, we have*

$$|e(B, C) - cbdn| \leq \lambda \sqrt{bc}\, n.$$

Proof. By Theorem 2.4,

$$\sum_{v \in C} (|N_B(v)| - bd)^2 \leq \sum_{v \in V} (|N_B(v)| - bd)^2 \leq \lambda^2 b(1-b)n.$$

Thus, by the Cauchy–Schwarz inequality,

$$\begin{aligned} |e(B, C) - cbdn| &\leq \sum_{v \in C} |N_B(v) - bd| \leq \sqrt{cn} \left(\sum_{v \in C} (|N_B(v)| - bd)^2 \right)^{1/2} \\ &\leq \sqrt{cn}\, \lambda \sqrt{b(1-b)n} \leq \lambda \sqrt{bc}\, n. \end{aligned}$$

■

The special case $B = C$ gives the following result. A slightly stronger estimate is proved in a similar way in Alon and Chung (1988).

Corollary 2.6. *Let $G = (V, E), d, n$, and λ be as in Theorem 2.4. Let B be an arbitrary set of bn vertices of G and let $e(B) = \frac{1}{2}e(B,B)$ be the number of edges in the induced subgraph of G on B. Then*

$$\left|e(B) - \tfrac{1}{2}b^2dn\right| \leq \tfrac{1}{2}\lambda bn.$$

A *walk of length l* in a graph G is a sequence $v_0, \ldots, v_l$ of vertices of G, where for each $1 \leq i \leq l$, $v_{i-1}v_i$ is an edge of G. Obviously, the total number of walks of length l in a d-regular graph on n vertices is precisely $n \cdot d^l$. Suppose, now, that C is a subset of, say, $n/2$ vertices of G. How many of these walks do not contain any vertex of C? In case G is disconnected, it may happen that half of these walks avoid C. However, as shown by Ajtai, Komlós, and Szemerédi (1987), there are many fewer such walks when all the eigenvalues of G but the largest have a small absolute value. This result and some of its extensions have several applications in theoretical computer science, as shown in the above-mentioned paper (see also Cohen and Wigderson [1990]). We conclude this section by stating and proving the result and one of its applications.

Theorem 2.7. *Let $G = (V, E)$ be a d-regular graph on n vertices, and suppose the absolute value of each of its eigenvalues but the first one is at most λ. Let C be a set of cn vertices of G. Then, for every l, the number of walks of length l in G that avoid C does not exceed $n(1-c)^{1/2}((1-c)d^2 + \lambda^2)^{l/2}$.*

Proof. Let $\{1, \ldots, n\}$ denote the set of vertices of G. Let A be the adjacency matrix of G and let P be the matrix of the projection on the set $V \backslash C$. Let x be a vector of n ones. We claim that for every $j \geq 0$, the ith coordinate of the vector $(PA)^j Px$ is precisely the number of walks of length j in G that avoid C and end in the vertex i. This is trivial for $j = 0$, and assuming it is true for $j - 1$, it is not too difficult to show it holds for j as well. Thus our objective is to find an upper bound for the sum of the coordinates of the vector $(PA)^l Px$, i.e., for the l_1-norm of this vector.

It is easier to bound its l_2-norm. To do so, we first claim that the l_2-norm of the operator PA is at most $((1-c)d^2 + \lambda^2)^{1/2}$. To prove this claim, suppose z is a real vector of l_2-norm $\|z\| = 1$. Put $z = u + v$, where u is a scalar multiple of the all-1 vector and v is a vector whose sum of coordinates is 0. Clearly, $\|u\|^2 + \|v\|^2 = \|z\|^2 = 1$. Now $Au = du$, and PAu is equal to Au in the coordinates corresponding to the vertices of $V \backslash C$ and is 0 in the other coordinates. Since all the coordinates of Au are equal, this implies that $\|PAu\| = \sqrt{1-c}\,d\|u\|$. On the other hand, since v is orthogonal to the eigenvector of the largest eigenvalue of A, $\|Av\| \leq \lambda\|v\|$, and hence $\|PAv\| \leq \lambda\|v\|$, since the norm of P is 1. By the triangle inequality, it follows that $\|PAz\| \leq \sqrt{1-c}\,d\|u\| + \lambda\|v\|$. This, by Cauchy–Schwarz, is at most

$((1-c)d^2+\lambda^2)^{1/2}(\|u\|^2+\|v\|^2)^{1/2}=((1-c)d^2+\lambda^2)^{1/2}$. Therefore the norm of PA is at most $((1-c)d^2+\lambda^2)^{1/2}$, as claimed.

Since the norm of Px is $\sqrt{1-c}\sqrt{n}$, we conclude that the l_2-norm of the vector $(PA)^l Px$ is at most $((1-c)d^2+\lambda^2)^{l/2}\sqrt{1-c}\sqrt{n}$. Thus, by Cauchy–Schwarz, the l_1-norm of this vector (which is precisely the number of walks that avoid C) is at most $n(1-c)^{1/2}((1-c)d^2+\lambda^2)^{l/2}$, completing the proof. ∎

The estimate given in the last theorem can be improved, by showing that the norm of $(PA)^j$ is, in fact, smaller than $\|PA\|^j$. For our purposes, however, the above estimate suffices.

A *randomly chosen walk* of length l in a graph G is a walk of length l in G chosen according to a uniform distribution among all walks of that length. Notice that if G is d-regular, such a walk can be chosen by choosing randomly its starting point v_0, and then by choosing, for each $1 \le i \le l$, v_i randomly among the d neighbors of v_{i-1}.

Corollary 2.8. *Let $G=(V,E)$, d, n, λ, C, and c be as in Theorem 2.7 and suppose*

$$((1-c)d^2+\lambda^2)^{1/2} \le \frac{d}{2^{1/4}}.$$

Then, for every l, the probability that a randomly chosen walk of length l in G avoids C is at most $2^{-l/4}$.

Proof. The number of walks of length l in G that avoid C is at most $n(1-c)^{1/2}((1-c)d^2+\lambda^2)^{l/2} \le nd^l 2^{-l/4}$, by Theorem 2.7. Since the total number of walks is nd^l, the desired result follows. ∎

The results above are useful for amplification of probabilities in randomized algorithms. Although such an amplification can be achieved for any Monte Carlo algorithm, we prefer, for simplicity, to consider one representative example: the primality testing algorithm of Rabin (1980).

For an odd number q, define two integers a and b by $q-1=2^a b$, where b is odd. An integer x, $1 \le x \le q-1$, is called a *witness* (for the nonprimality of q) if for the sequence $x_0,\dots,x_a$ defined by $x_0=x^b(\bmod q)$ and $x_i = x_{i-1}^2(\bmod q)$ for $1 \le i \le a$, either $x_a \ne 1$ or there is an i such that $x_i \ne -1,1$ and $x_{i+1}=1$. One can show that if q is a prime, then there are no such witnesses for q, whereas if q is an odd nonprime, then at least half of the numbers between 1 and $q-1$ are witnesses for q. (In fact, at least 3/4 are witnesses, as shown by Rabin). This suggests the following randomized algorithm for testing if an odd number q is a prime.

Choose, randomly, an integer x between 1 and $q-1$ and check if it is a witness. If it is, report that q is not a prime. Otherwise, report that q is a prime.

Observe that in case q is a prime, the algorithm certainly reports it is a prime, whereas in case q is not a prime, the probability that the algorithm makes a mistake and reports it as a prime is at most 1/2. What if we wish to reduce the probability of making such a mistake? Clearly, we can simply repeat the algorithm. If we repeat it l independent times, then the probability of making an error (i.e., reporting a nonprime as a prime) decreases to $1/2^l$. However, the number of random bits required for this procedure is $l \cdot \log(q-1)$.

Suppose we wish to use less random bits. By applying the properties of a randomly chosen walk on an appropriate graph, proved in the last two results, we can obtain the same estimate for the error probability by using only $(1 + o(1))\log(q-1) + O(l)$ random bits. This is done as follows.

Let G be a d-regular graph with $q-1$ vertices, labeled by all integers between 1 and $q-1$. Suppose G has no eigenvalue but the first one, whose absolute value exceeds λ, and suppose that

$$\left(\frac{1}{2}d^2 + \lambda^2\right)^{1/2} \leq \frac{d}{2^{1/4}}. \tag{1}$$

Now choose randomly a walk of length $4l$ in the graph G, and check, for each of the numbers labeling its vertices, if it is a witness. If q is a nonprime, then at least half of the vertices of G are labeled by witnesses. Hence, by Corollary 2.8 and by (1), the probability that no witness is on the walk is at most $2^{-4l/4} = 2^{-l}$. Thus we obtain the same reduction in the error probability as the one obtained by choosing l independent witnesses. Let us estimate the number of random bits required for choosing such a random walk.

The known constructions of expanders given by Lubotzky, Phillips, and Sarnak (1986) or by Margulis (1988) give explicit families of graphs with degree d and with $\lambda \leq 2\sqrt{d-1}$, for each $d = p+1$, where p is a prime congruent to 1 modulo 4. (We note that these graphs will not have exactly $q-1$ vertices but this does not cause any real problem as we can take a graph with n vertices, where $q \leq n \leq (1+o(1))q$, and label its ith vertex by $i \pmod{q-1}$. In this case, the number of vertices labeled by witnesses would still be at least $(1/2 + o(1))n$.) One can easily check that, e.g., $d = 30$ and $\lambda = 2\sqrt{29}$ satisfy (1), and thus we can use a 30-regular graph. The number of random bits required for choosing a random walk of length $4l$ in it is less than $(1+o(1))\log q + 20l$, much less than the $l\log(q-1)$ bits that are needed in the repetition procedure.

3. QUASI-RANDOM GRAPHS

In this section, we describe several pseudo-random properties of graphs which, somewhat surprisingly, turn out to be all equivalent. All the properties are ones satisfied, almost surely, by a random graph in which every edge is chosen, independently, with probability 1/2. The equivalence between some of these properties were first proved by several authors—see Thomason (1987), Frankl, Rödl, and Wilson (1988), and Alon and Chung (1988)—but the first paper in which all of them (and some others) appear together is the one by Chung, Graham, and Wilson (1989). Our presentation here follows that paper, although in order to simplify the presentation, we consider only the case of regular graphs.

We first need some notation. For two graphs G and H, let $N_G^*(H)$ be the number of labeled occurrences of H as an induced subgraph of G (i.e., the number of adjacency-preserving injections $f : V(H) \mapsto V(G)$ whose image is the set of vertices of an induced copy of H in G). Similarly, $N_G(H)$ denotes the number of labeled copies of H as a (not necessarily induced) subgraph of G. Note that $N_G(H) = \sum_L N_G^*(L)$, where L ranges over all graphs on the set of vertices of H obtained from H by adding to it a (possibly empty) set of edges.

Throughout this section, G always denotes a graph with n vertices. We denote the eigenvalues of its adjacency matrix (taken with multiplicities) by $\lambda_1, \ldots, \lambda_n$, where $|\lambda_1| \geq \cdots \geq |\lambda_n|$. (Since we consider in this section only the eigenvalues of G, we simply write λ_1 and not $\lambda_1(G)$.) Recall also the following notation, used in the previous section: for a vertex v of G, $N(v)$ denotes the set of its neighbors in G. If S is a set of vertices of G, $e(S)$ denotes the number of edges in the induced subgraph of G on S. If B and C are two (not necessarily disjoint) subsets of vertices of G, $e(B,C)$ denotes the number of ordered pairs (b,c) where $b \in B$, $c \in C$, and bc is an edge of G. Thus $e(S) = \frac{1}{2}e(S,S)$.

We can now state the pseudo-random properties considered here. All the properties refer to a graph $G = (V,E)$ with n vertices. Throughout the section, we use the $o(\cdot)$-notation, without mentioning the precise behavior of each $o(\cdot)$. Thus occurrences of two $o(1)$, say, need not mean that both are identical and only mean that if we consider a family of graphs G and let their number of vertices n tend to infinity, then each $o(1)$ tends to 0.

Property $P_1(s)$. *For every graph* $H(s)$ *on* s *vertices,* $N_G^*(H(s)) = (1+o(1)) n^s 2^{-\binom{s}{2}}$.

Property P_2. *For the cycle* $C(4)$ *with* 4 *vertices,* $N_G(C(4)) \leq (1+o(1))(n/2)^4$.

Property P_3. $|\lambda_2| = o(n)$.

Property P_4. *For every set* S *of vertices of* G, $e(S) = 1/4|S|^2 + o(n^2)$.

Property P_5. *For every two sets of vertices* B *and* C, $e(B,C) = 1/2|B||C| + o(n^2)$.

Property P_6. $\sum_{u,v \in V} \big| |N(u) \cap N(v)| - n/4 \big| = o(n^3)$.

It is easy to check that all the properties above are satisfied, almost surely, by a random graph on n vertices. In this section, we show that all these properties are equivalent for a regular graph with n vertices and degree of regularity about $n/2$. The fact that the innocent-looking property P_2 is strong enough to imply for such graphs $P_1(s)$ for every $s \geq 1$ is one of the interesting special cases of this result.

Graphs that satisfy any (and thus all) of the properties above are called *quasi-random*. As noted above, the assumption that G is regular can be dropped (at the expense of slightly modifying property P_2 and slightly complicating the proofs).

Theorem 3.1. *Let G be a d-regular graph on n vertices, where $d = (1/2 + o(1))n$. If G satisfies any one of the seven properties $P_1(4)$, $P_1(s)$ for all $s \geq 1$, P_2, P_3, P_4, P_5, P_6, then it satisfies all seven.*

Proof. We show that

$$\begin{aligned} &P_1(4) \Rightarrow P_2 \Rightarrow P_3 \Rightarrow P_4 \Rightarrow P_5 \\ &\Rightarrow P_6 \Rightarrow P_1(s) \qquad \text{for all} \quad s \geq 1 \ (\Rightarrow P_1(4)). \end{aligned}$$

1. $P_1(4) \Rightarrow P_2$. Suppose G satisfies $P_1(4)$. Then $N_G(C(4)) = \sum_L N_G^*(L)$, as L ranges over the four labeled graphs obtained from a labeled $C(4)$ by adding to it a (possibly empty) set of edges. Since G satisfies $P_1(4)$, $N_G^*(L) = (1+o(1))n^4 2^{-6}$ for each of these graphs L and hence $N_G(C(4)) = (1+o(1)) n^4 2^{-4}$, showing that G satisfies P_2.

2. $P_2 \Rightarrow P_3$. Suppose G satisfies P_2 and let A be its adjacency matrix. The trace of A^4 is precisely $\sum_{i=1}^n \lambda_i^4$. On the other hand, it is easy to see that this trace is precisely the number of (labeled) closed walks of length 4 in G, i.e., the number of sequences $v_0, v_1, v_2, v_3, v_4 = v_0$ of vertices of G such that $v_i v_{i+1}$ is an edge for each $0 \leq i \leq 3$. This number is $N_G((C(4))$ plus the number of such sequences in which $v_2 = v_0$, which is nd^2, plus the number of such sequences in which $v_2 \neq v_0$ and $v_3 = v_1$, which is $nd(d-1)$. Thus

$$\begin{aligned} \sum_{i=1}^n \lambda_i^4 &= d^4 + \sum_{i=2}^n \lambda_i^4 = (1+o(1)) \left(\frac{n}{2}\right)^4 + \sum_{i=2}^n \lambda_i^4 \\ &= N_G(C(4)) + O(n^3) = (1+o(1)) \left(\frac{n}{2}\right)^4. \end{aligned}$$

It follows that $\sum_{i=2}^n \lambda_i^4 = o(n^4)$, and hence that $|\lambda_2| = o(n)$, as needed.

3. $P_3 \Rightarrow P_4$. This is an immediate consequence of Corollary 2.6.

4. $P_4 \Rightarrow P_5$. Suppose G satisfies P_4. We first claim that it satisfies property P_5 for disjoint sets of vertices B and C. Indeed, if B and C are disjoint, then

$$\begin{aligned} e(B,C) &= e(B \cup C) - e(B) - e(C) \\ &= \tfrac{1}{4}(|B| + |C|)^2 - \tfrac{1}{4}|B|^2 - \tfrac{1}{4}|C|^2 + o(n^2) \\ &= \tfrac{1}{2}|B|\,|C| + o(n^2), \end{aligned}$$

proving the claim.

In case B and C are not disjoint, we have

$$e(B,C) = e(B \backslash C, C \backslash B) + e(B \cap C, C \backslash B) + e(B \cap C, B \backslash C) + 2e(B \cap C).$$

Put $|B| = b$, $|C| = c$, $|B \cap C| = x$. By the above expression for $e(B,C)$ and by the fact that G satisfies P_4 and P_5 for disjoint B and C, we get

$$\begin{aligned} e(B,C) &= \tfrac{1}{2}(b-x)(c-x) + \tfrac{1}{2}x(c-x) + \tfrac{1}{2}x(b-x) + \tfrac{2}{4}x^2 + o(n^2) \\ &= \tfrac{1}{2}bc + o(n^2) = \tfrac{1}{2}|B|\,|C| + o(n^2), \end{aligned}$$

showing that G satisfies P_5.

5. $P_5 \Rightarrow P_6$. Suppose that G satisfies P_5 and recall that G is d-regular, where $d = (1/2 + o(1))n$. Let v be a fixed vertex of G, and let us estimate the sum

$$\sum_{u \in V, u \neq v} \left| |N(u) \cap N(v)| - \frac{n}{4} \right|.$$

Define

$$B_1 = \left\{ u \in V, u \neq v : |N(u) \cap N(v)| \geq \frac{n}{4} \right\},$$

and similarly,

$$B_2 = \left\{ u \in V, u \neq v : |N(u) \cap N(v)| < \frac{n}{4} \right\}.$$

Let C be the set of all neighbors of v in G. Observe that

$$\begin{aligned} \sum_{u \in B_1} \left| |N(u) \cap N(v)| - \frac{n}{4} \right| &= \sum_{u \in B_1} |N(u) \cap N(v)| - |B_1| \frac{n}{4} \\ &= e(B_1, C) - |B_1| \frac{n}{4}. \end{aligned}$$

Since G satisfies P_5 and since $d = (1/2 + o(1))n$, the last difference is $1/2|B_1|d + o(n^2) - |B_1|n/4 = o(n^2)$.

A similar argument implies that

$$\sum_{u \in B_2} \left| |N(u) \cap N(v)| - \frac{n}{4} \right| = o(n^2).$$

It follows that for every vertex v of G,

$$\sum_{u \in V, u \neq v} \left| |N(u) \cap N(v)| - \frac{n}{4} \right| = o(n^2),$$

and by summing over all vertices v, we conclude that G satisfies property P_6.

6. $P_6 \Rightarrow P_1(s)$ **for all** $s \geq 1$. Suppose $G = (V, E)$ satisfies P_6. For any two distinct vertices u and v of G, let $a(u,v)$ be 1 if $uv \in E$ and 0 otherwise. Also, define $s(u,v) = |\{w \in V : a(u,w) = a(v,w)\}|$. Since G is $d = (1/2 + o(1))n$-regular, $s(u,v) = 2|N(u) \cap N(v)| + n - 2d = 2|N(u) \cap N(v)| + o(n)$. Therefore, the fact that G satisfies P_6 implies that

$$\sum_{u,v \in V} \left| s(u,v) - \frac{n}{2} \right| = o(n^3). \tag{2}$$

Let $H = H(s)$ be an arbitrary fixed graph on s vertices, and put $N_s = N_G^*(H(s))$. We must show that

$$N_s = (1 + o(1))n^s 2^{-\binom{s}{2}}.$$

Denote the vertex set of $H(s)$ by $\{v_1, \ldots, v_s\}$. For each $1 \leq r \leq s$, put $V_r = \{v_1, \ldots, v_r\}$, and let $H(r)$ be the induced subgraph of H on V_r. We prove, by induction on r, that for $N_r = N_G^*(H(r))$,

$$N_r = (1 + o(1))n_{(r)} 2^{-\binom{r}{2}}, \tag{3}$$

where $n_{(r)} = n(n-1)\ldots(n-r+1)$.

This is trivial for $r = 1$. Assuming it holds for r, where $1 \leq r < s$, we prove it for $r+1$. For a vector $\alpha = (\alpha_1, \ldots, \alpha_r)$ of distinct vertices of G, and for a vector $\epsilon = (\epsilon_1, \ldots, \epsilon_r)$ of $\{0,1\}$-entries, define

$$f_r(\alpha, \epsilon) = \left| \{v \in V : v \neq \alpha_1, \ldots, \alpha_r \text{ and } a(v, \alpha_j) = \epsilon_j \text{ for all } 1 \leq j \leq r\} \right|.$$

Clearly, N_{r+1} is the sum of the N_r quantities $f_r(\alpha, \epsilon)$ in which $\epsilon_j = a(v_{r+1}, v_j)$ and α ranges over all N_r induced copies of $H(r)$ in G.

Observe that altogether there are precisely $n_{(r)} 2^r$ quantities $f_r(\alpha, \epsilon)$. It is convenient to view $f_r(\alpha, \epsilon)$ as a random variable defined on a sample space of $n_{(r)} 2^r$ points, each having an equal probability. To complete the proof, we compute the expectation and the variance of this random variable. We show that the variance is so small that most of the quantities $f_r(\alpha, \epsilon)$ are very close to the expectation, and thus obtain a sufficiently accurate estimate for N_{r+1}, which is the sum of N_r such quantities.

We start with the simple computation of the expectation $E(f_r)$ of $f_r(\alpha, \epsilon)$. We have

$$E(f_r) = \frac{1}{n_{(r)} 2^r} \sum_{\alpha, \epsilon} f_r(\alpha, \epsilon) = \frac{1}{n_{(r)} 2^r} \sum_{\alpha} \sum_{\epsilon} f_r(\alpha, \epsilon) = \frac{1}{n_{(r)} 2^r} \sum_{\alpha} (n - r) = \frac{n-r}{2^r},$$

where we used the fact that every vertex $v \neq \alpha_1, \ldots, \alpha_r$ defines ϵ uniquely.

Next, we estimate the quantity S_r, defined by

$$S_r = \sum_{\alpha,\epsilon} f_r(\alpha,\epsilon)(f_r(\alpha,\epsilon)-1).$$

We claim that

$$S_r = \sum_{u \neq v} s(u,v)_{(r)}. \tag{4}$$

To prove this claim, observe that S_r can be interpreted as the number of ordered triples $(\alpha,\epsilon,(u,v))$, where $\alpha = (\alpha_1,\ldots,\alpha_r)$ is an ordered set of r distinct vertices of G, $\epsilon = (\epsilon_1,\ldots,\epsilon_r)$ is a binary vector of length r, and u,v is an ordered pair of additional vertices of G so that

$$a(u,\alpha_k) = a(v,\alpha_k) = \epsilon_k \qquad \text{for all} \quad 1 \leq k \leq r.$$

For each fixed α and ϵ, there are precisely $f_r(\alpha,\epsilon)(f_r(\alpha,\epsilon)-1)$ choices for the pair (u,v) and hence S_r counts the number of these triples.

Now let us compute this number by first choosing u and v. Once u,v are chosen, the additional vertices $\alpha_1,\ldots,\alpha_r$ must all belong to the set $\{w \in V : a(u,w) = a(v,w)\}$. Since the cardinality of this set is $s(u,v)$, it follows that there are $s(u,v)_{(r)}$ choices for $\alpha_1,\ldots,\alpha_r$. Once these are chosen, the vector ϵ is determined and thus (4) follows.

We next claim that (2) implies

$$\sum_{u \neq v} s(u,v)_{(r)} = (1+o(1))n^{r+2}2^{-r}. \tag{5}$$

To prove this claim, define $\epsilon_{uv} = s(u,v) - n/2$. Observe that by (2), $\sum_{u \neq v} |\epsilon_{uv}| = o(n^3)$, and that $|\epsilon_{uv}| \leq n/2 \leq n$ for each u,v. Hence, for every fixed $a \geq 1$,

$$\sum_{u \neq v} |\epsilon_{uv}|^a \leq n^{a-1} \sum_{u \neq v} |\epsilon_{uv}| = o(n^{a+2}).$$

This implies that

$$\begin{aligned}
\sum_{u \neq v} s(u,v)_{(r)} &= \sum_{u \neq v} \left(\frac{n}{2} + \epsilon_{uv}\right)_{(r)} \\
&= \sum_{k=0}^{r} \sum_{u \neq v} c_k \left(\frac{n}{2}\right)^k \epsilon_{uv}^{r-k} \qquad \text{(for appropriate constants } c_k\text{)} \\
&= \left(\frac{n}{2}\right)^r n_{(2)} + \sum_{k=0}^{r-1} \sum_{u \neq v} c_k \left(\frac{n}{2}\right)^k \epsilon_{uv}^{r-k}
\end{aligned}$$

$$\leq \left(\frac{n}{2}\right)^r n_{(2)} + \sum_{k=0}^{r-1} \sum_{u \neq v} |c_k| n^k |\epsilon_{uv}|^{r-k}$$

$$\leq n^{r+2} 2^{-r} + c \sum_{k=0}^{r-1} n^k \sum_{u \neq v} |\epsilon_{uv}|^{r-k}$$

(for an appropriate constant c)

$$\leq n^{r+2} 2^{-r} + c \sum_{k=0}^{r-1} n^k \cdot o(n^{r-k+2})$$

$$= n^{r+2} 2^{-r} (1 + o(1)),$$

implying (5).

By (4) and (5),

$$S_r = (1 + o(1)) n^{r+2} 2^{-r}.$$

Therefore

$$\sum_{\alpha,\epsilon} (f_r(\alpha,\epsilon) - E(f_r))^2 = \sum_{\alpha,\epsilon} f_r^2(\alpha,\epsilon) - \sum_{\alpha,\epsilon} E(f_r)^2$$

$$= \sum_{\alpha,\epsilon} (f_r^2(\alpha,\epsilon) - f_r(\alpha,\epsilon))$$

$$+ \sum_{\alpha,\epsilon} f_r(\alpha,\epsilon) - n_{(r)} 2^r (n-r)^2 2^{-2r}$$

$$= S_r + n_{(r)} 2^r E(f_r) - n_{(r)} 2^r (n-r)^2 2^{-2r}$$

$$= S_r + n_{(r+1)} - n_{(r)} 2^r (n-r)^2 2^{-2r} = o(n^{r+2}).$$

Recall that N_{r+1} is the summation of N_r quantities of the form $f_r(\alpha,\epsilon)$. Thus

$$|N_{r+1} - N_r E(f_r)|^2 = \left| \sum_{N_r \text{ terms}} (f_r(\alpha,\epsilon) - E(f_r)) \right|^2.$$

By Cauchy–Schwarz, the last expression is at most

$$N_r \sum_{N_r \text{ terms}} (f_r(\alpha,\epsilon) - E(f_r))^2 \leq N_r \sum_{\alpha,\epsilon} (f_r(\alpha,\epsilon) - E(f_r))^2$$

$$= N_r \cdot o(n^{r+2}) = o(n^{2r+2}).$$

It follows that

$$|N_{r+1} - N_r E(f_r)| = o(n^{r+1}),$$

and hence, by the induction hypothesis,

$$\begin{aligned} N_{r+1} &= N_r E(f_r) + o(n^{r+1}) \\ &= (1+o(1)) n_{(r)} 2^{-\binom{r}{2}} \cdot (n-r) 2^{-r} + o(n^{r+1}) \\ &= (1+o(1)) n_{(r+1)} 2^{-\binom{r+1}{2}}. \end{aligned}$$

This completes the proof of the induction step and establishes Theorem 3.1. ■

There are many examples of families of quasi-random graphs. The most widely used is probably the family of Paley graphs Q_p defined as follows. For a prime p congruent to 1 modulo 4, let G_p be the graph whose vertices are the integers $0, 1, 2, \ldots, p-1$ in which i and j are adjacent if and only if $i-j$ is a quadratic residue modulo p. The graphs G_p, which are the undirected analogs of the quadratic residue tournaments discussed in Section 1, are $(p-1)/2$-regular. For any two distinct vertices i and j of G_p, the number of vertices k that are either adjacent to both i and j or nonadjacent to both is precisely the number of times the quotient $(k-i)/(k-j)$ is a quadratic residue. As k ranges over all numbers between 0 and $p-1$ but i and j, this quotient ranges over all numbers but 1 and 0 and hence it is a quadratic residue precisely $1/2(p-1)-1$ times. (This is essentially the same assertion as that of the first fact given in the proof of Theorem 1.1.) We have thus shown that for every two vertices i and j of G_p, $s(i,j) = (p-3)/2$, and this, together with the fact that G_p is $(p-1)/2$-regular, easily implies that it satisfies property P_6. Therefore it is quasi-random. As is the case with the quadratic residue tournaments, G_p satisfies, in fact, some stronger pseudo-random properties that are not satisfied by every quasi-random graph, and that can be proved by applying Weil's theorem.

4. THE RÖDL NIBBLE

For $2 \le l < k < n$, let $M(n,k,l)$, the covering number, denote the minimal size of a family $\mathcal{K}$ of k-element subsets of $\{1, \ldots, n\}$ having the property that every l-element set is contained in at least one $A \in \mathcal{K}$. Clearly, $M(n,k,l) \ge \binom{n}{l}/\binom{k}{l}$, since each k-set covers $\binom{k}{l}$ l-sets and every l-set must be covered. Equality holds if and only if the family $\mathcal{K}$ has the property that every l-set is contained in exactly one $A \in \mathcal{K}$. This is called an (n,k,l) tactical configuration (or block design). For example, $(n,3,2)$ tactical configurations are better known as Steiner triple systems. The question of the existence of tactical configurations is a central one for combinatorics, but one for which probabilistic methods (at least so far!) play little role. In 1963, Paul Erdős and Haim Hanani conjectured that for and fixed $2 \le l < k$

$$\lim_{n \to \infty} \frac{M(n,k,l)}{\binom{n}{l} / \binom{k}{l}} = 1.$$

Their conjecture was, roughly, that one can get asymptotically close to a tactical configuration. While this conjecture seemed ideal for a probabilistic analysis, it was a full generation before Vojtech Rödl (1985) found the proof, which we outline in this section. (One may similarly define the packing number $m(n,k,l)$ as the maximal size of a family $\mathcal{K}$ of k-element subsets of $\{1,\ldots,n\}$ having the property that every l-element set is contained in at most one $A \in \mathcal{K}$. Erdős and Hanani noticed from elementary arguments that

$$\lim_{n\to\infty} \frac{M(n,k,l)}{\binom{n}{l}/\binom{k}{l}} = 1 \iff \lim_{n\to\infty} \frac{m(n,k,l)}{\binom{n}{l}/\binom{k}{l}} = 1.$$

While Rödl's result may be formulated in terms of either packing or covering, here we deal only with the covering problem.)

The first key to the Rödl argument is to randomly select a *small* number of k-sets, what we call the Rödl nibble. Let $\epsilon > 0$ be fixed but very small. Select k-sets randomly with appropriate probability so that an l-set is, on average, covered ϵ times. The Poisson paradigm applies and each l-set has probability $e^{-\epsilon}$ of not being covered. A proportion $1-e^{-\epsilon} \sim \epsilon - \epsilon^2/2$ of l-sets are covered, while if no l-set would have been covered, twice a proportion ϵ would have been covered. The "efficiency" of the covering is $(1-e^{-\epsilon})/\epsilon$, which is nearly 1. Now let G_1 be the family of l-sets that have not been covered, regarded as an l-uniform hypergraph. From among the k-sets that are cliques in G_1 (i.e., all of whose l-element subsets are not yet covered), we again select a random subfamily so that an l-set of G_1 is, on average, covered ϵ times. Note, critically, that these new k-sets cannot cover any l-set that was already covered in the first random selection. Again, we shall see, the efficiency is near 1. We iterate with $G_2, G_3, \ldots, G_{t-1}$. We stop the procedure with a small proportion of l-sets uncovered. At this point, for each uncovered l-set, we select some k-set covering it. While this last part is terribly inefficient (we want each k-set to cover $\binom{k}{l}$ new l-sets but here each covers only one), it is acceptable, since it only involves a negligible proportion of the k-sets. The total efficiency is then near 1.

The l-graph G_1 has density $\rho = e^{-\epsilon}$, but it is certainly not a random l-graph with density ρ. The second key to the Rödl argument is to show that G_1 (and later iterates $G_2, \ldots, G_{t-1}$) retain essential properties of random graphs, that they are quasi-random in an appropriate sense. One shows that given a quasi-random G_i, the random selection of k-cliques that covers each l-set on average ϵ times will leave a graph G_{i+1} of uncovered l-sets, which is itself quasi-random. The notion of quasi-random hypergraphs has been developed by Fan Chung and Ronald Graham (1990). In retrospect, it was the original proof of Rödl that provided early motivation for the placing of quasi-random graphs and quasi-random hypergraphs on solid footing. His result has been extended in several ways; see, e.g., Pippenger and Spencer (1989) and Kahn (1990).

We have seen in Section 3 that quasi-random graphs require only an asymptotically appropriate count of the number of subgraphs of each type on four vertices. For l-uniform hypergraphs, Chung and Graham showed that the asymptotically appropriate count of the number of subgraphs of each type on $2l$ vertices was sufficient. Let A be a family of l-element subsets of $\{1,\ldots,2l\}$, a subgraph type. For any G (all graphs are henceforth l-uniform hypergraphs), let $N_A(G)$ denote the number of $2l$-tuples $x=(x_1,\ldots,x_{2l})$ of vertices of G so that $\{x_{i_1},\ldots,x_{i_l}\} \in G$ whenever $\{i_1,\ldots,i_l\} \in A$. G is quasi-random with density ρ if for all A, $N_A(G) \sim n^{2l}\rho^{|A|}$. This is naturally the case if G is random with edge probability ρ. (The count $N_A(G)$ gives the subgraphs containing A and possibly more edges, but the condition that these counts are correct for all A gives, via inclusion–exclusion, that for all A there are an appropriate number of subgraphs with exactly the edges of A.) We shall use that this condition implies

- G has $\sim \binom{n}{k}\rho^{\binom{k}{l}}$ k-cliques,
- For almost every edge e of G, e lies in $\sim \binom{n}{k-l}\rho^{\binom{k}{l}-1}$ k-cliques.

We omit the proofs. The $\rho = 1/2$ case is in Chung and Graham (1990). With this definition, we give a theorem that allows the iteration to work.

Theorem 4.1. *Let $0<\epsilon, \rho<1$, l be fixed. Let G be a quasi-random l-uniform hypergraph with density ρ. Then there exists a set $\mathcal{K}$ of k-cliques of G of size $\sim \epsilon\rho\binom{n}{l}/\binom{k}{l}$ so that, letting H denote the graph of edges of G not contained in any $K \in \mathcal{K}$, H is quasi-random with density $\rho e^{-\epsilon}$.*

Proof. Place each k-clique of G independently in $\mathcal{K}$ with probability

$$q = \frac{\epsilon\rho\binom{n}{l}/\binom{k}{l}}{\binom{n}{k}\rho^{\binom{k}{l}}}$$

(which is less than 1).

As the denominator of q is asymptotically, from quasi-randomness of G, the total number of k-cliques of G, $|\mathcal{K}|$ will have binomial distribution with mean the numerator of q. By Chebyschev's inequality (or the much stronger results of Appendix A), $\mathcal{K}$ will almost surely be of size $\sim \epsilon\rho\binom{n}{l}/\binom{k}{l}$. (Were $\mathcal{K}$ nonoverlapping, $\sim \epsilon\rho\binom{n}{l}$ edges of G would be covered, a proportion ϵ of the $\rho\binom{n}{l}$ edges of G.)

Now fix a family A of l-element subsets of $\{1,\ldots,2l\}$, say $|A| = a$. For any $2l$-tuple $x=(x_1,\ldots,x_{2l})$ of vertices of G and any $e=\{i_1,\ldots,i_l\} \in A$, let x_e, for notational convenience, denote $\{x_{i_1},\ldots,x_{i_l}\}$. Call x good for G if $x_e \in G$ for all $e \in A$ and call x good for H if $x_e \in H$ for all $e \in A$. Let $X = N_A(H)$, the number of x good for H. Call an edge of G normal if it lies in $\sim \alpha n^{k-l}$

k-cliques, where we set $\alpha = \rho^{\binom{k}{l}-1}/(k-l)!$ for notational convenience. Otherwise, call the edge abnormal. When x is good for G, call x normal if all x_e, $e \in A$ are normal, otherwise call x abnormal. Quasi-randomness of G gives that there are $\sim n^{2l}\rho^a$ x that are good for G. Quasi-randomness of G also gives that only $o(n^l)$ edges of G are abnormal, and so there are only $o(n^{2l})$ abnormal x and thus $\sim n^{2l}\rho^a$ normal x good for G.

Let I_x be the indicator random variable for x being good for H. Then $X = \sum I_x$, the sum over x good for G. Suppose x is normal. For each $e \in A$, x_e lies in $\sim \alpha n^{k-l}$ k-cliques. Any two $x_e, x_{e'}$ can be commonly contained in only $O(n^{k-l-1}) = o(n^{k-l})$ k-cliques. Thus there are $\sim a\alpha n^{k-l}$ k-cliques containing some x_e, $e \in A$. Hence

$$E[I_x] = (1-q)^{a\alpha n^{k-l}(1+o(1))} = \exp\left(-qa\alpha n^{k-l}(1+o(1))\right)$$
$$= \exp(-a\epsilon(1+o(1))) = \exp(-a\epsilon) + o(1).$$

When x is abnormal, $E[I_x] \leq 1$, so these contribute $o(n^{2l})$ to $E[X]$. By linearity of expectation,

$$E[X] \sim n^{2l}(\rho e^{-\epsilon})^a.$$

We expand the variance as in Chapter 4:

$$\text{var}[X] \leq E[X] + \sum \text{cov}[I_x, I_y],$$

the sum over distinct x, y good for G. There are only $o(n^{4l})$ terms in which either x or y is abnormal and only $O(n^{4l-1}) = o(n^{4l})$ terms in which x, y overlap. As $\text{cov}[I_x, I_y] \leq E[I_x I_y] \leq 1$ always, these contribute $o(n^{4l})$ and $\text{var}[X] = o(n^{4l}) + \sum^* \text{cov}[I_x, I_y]$, where $\sum^*$ ranges over those x, y that are both good for G, both normal, and nonoverlapping. For such x, y, there are $\sim a\alpha n^{k-l}$ k-cliques covering some x_e and $\sim a\alpha n^{k-l}$ k-cliques covering some y_e. There are only $O(n^{k-2l}) = o(n^{k-l})$ k-cliques covering both an x_e and a $y_{e'}$. Thus there are a total of $\sim 2a\alpha n^{k-l}$ k-cliques containing either an x_e or a $y_{e'}$. Thus

$$E[I_x I_y] = (1-q)^{2a\alpha n^{k-l}(1+o(1))} = \exp\left(-2qa\alpha n^{k-l}(1+o(1))\right)$$
$$= \exp(-2a\epsilon(1+o(1))) = \exp(-2a\epsilon) + o(1)$$

and therefore

$$\text{cov}[I_x I_y] = E[I_x I_y] - E[I_x]E[I_y]$$
$$= e^{-2a\epsilon} + o(1) - (e^{-a\epsilon} + o(1))(e^{-a\epsilon} + o(1)) = o(1).$$

As there are $O(n^{4l})$ such terms

$$\text{var}[X] = o(n^{4l}) = o(E[X]^2),$$

and so, by Chebyschev's inequality, $X \sim n^{2l}(\rho e^{-\epsilon})^a$ almost surely.

This analysis holds for each of the finite number of types A. Hence almost surely it holds for all A simultaneously and also $|\mathcal{K}|$ is as desired. Thus there exists a specific $\mathcal{K}$ satisfying the theorem. ■

Theorem 4.2 (Rödl). *For k,l fixed,*

$$M(n,k,l) \le \binom{n}{l} \Big/ \binom{k}{l} (1 + o(1)).$$

Proof. Let $\epsilon > 0$, $t \in N$, be, for the moment, arbitrary. Let G_0 be the complete l-graph on n vertices. We define G_i inductively. For $0 \le i < t$, assume inductively that G_i is quasi-random with density $e^{-i\epsilon}$. (This surely holds for $i = 0$.) Let $\mathcal{K}_i$, as given by Theorem 4.1, be a family of cliques of G_i of size $\sim \epsilon e^{-i\epsilon}\binom{n}{l}/\binom{k}{l}$ and let G_{i+1} be the edges of G_i not lying in any of the cliques of $\mathcal{K}_i$. (Note that we are here applying Theorem 4.1 an arbitrary but fixed number of times t.) Let $\mathcal{K}_\infty$ consist of one k-set (whether clique or not) for every edge of G_t, a k-set covering the edge, so that

$$|\mathcal{K}_\infty| \le |G_t| \sim \binom{n}{l} e^{-\epsilon t}.$$

Set

$$\mathcal{K} = \mathcal{K}_0 \cup \cdots \cup \mathcal{K}_{t-1} \cup \mathcal{K}_\infty$$

so that $\mathcal{K}$ is a family of k-sets covering every l-element subset of $\{1,\ldots,n\}$. We calculate

$$|\mathcal{K}| \le \left[\binom{n}{l} \Big/ \binom{k}{l}\right] F(\epsilon,t) + o(n^l),$$

where

$$F(\epsilon,t) = \sum_{i=0}^{t-1} \epsilon e^{-i\epsilon} + \binom{k}{l} e^{-\epsilon t}.$$

For ϵ fixed,

$$\lim_{t\to\infty} F(\epsilon,t) = \frac{\epsilon}{1-e^{-\epsilon}}.$$

This reflects the notion that if we iterate the Rödl nibble long enough, the inefficiency at the end is negligible. We further note

$$\lim_{\epsilon\to 0} \frac{\epsilon}{1-e^{-\epsilon}} = 1.$$

This reflects the notion that if the Rödl nibble is small enough, the k-sets chosen will have negligible overlap.

Let $\delta > 0$ be arbitrarily small. There exists $\epsilon > 0$ with $\epsilon/(1-e^{-\epsilon}) < 1 + \delta/4$. Then there exists $t \in N$ with $F(\epsilon,t) < 1 + \delta/4$. Now fix those ϵ, t. For n sufficiently large, the $o(n^l)$ term in bounding $|\mathcal{K}|$ is less than $\delta\binom{n}{l}/4\binom{k}{l}$. For such n,

$$M(n,k,l) \leq |\mathcal{K}| \leq \left[\binom{n}{l} \Big/ \binom{k}{l}\right](1+\delta),$$

as desired. ■

The Probabilistic Lens: Random Walks

A *vertex-transitive* graph is a graph $G = (V,E)$ such that for any two vertices $u, v \in V$, there is an automorphism of G that maps u into v. A *random walk* of length l in G starting at a vertex v is a randomly chosen sequence $v = v_0, v_1, \ldots, v_l$, where each v_{i+1} is chosen, randomly and independently, among the neighbors of v_i $(0 \leq i < l)$.

The following theorem states that for every vertex-transitive graph G, the probability that a random walk of even length in G ends at its starting point is at least as big as the probability that it ends at any other vertex. Note that the proof requires almost no computation. We note also that the result does not hold for general regular graphs, and the vertex-transitivity assumption is necessary.

Theorem. *Let $G = (V,E)$ be a vertex-transitive graph. For an integer k and for two (not neccessarily distinct) vertices u, v of G, let $P^k(u,v)$ denote the probability that a random walk of length k starting at u ends at v. Then, for every integer k and for every two vertices $u, v \in V$,*

$$P^{2k}(u,u) \geq P^{2k}(u,v).$$

Proof. We need the following simple inequality, sometimes attributed to Chebyschev.

Claim. *For every sequence $(a_1, \ldots, a_n)$ of n reals and for any permutation π of $\{1, \ldots, n\}$,*

$$\sum_{i=1}^{n} a_i a_{\pi(i)} \leq \sum_{i=1}^{n} a_i^2.$$

Proof. The inequality follows immediately from the fact that

$$\sum_{i=1}^{n} a_i^2 - \sum_{i=1}^{n} a_i a_{\pi(i)} = \frac{1}{2} \sum_{i=1}^{n} (a_i - a_{\pi(i)})^2 \geq 0. \qquad \blacksquare$$

Proof of Theorem, Continued. Consider, now, a random walk of length $2k$ starting at u. By summing over all the possibilities of the vertex the walk reaches after k steps, we conclude that for every vertex v,

$$P^{2k}(u,v) = \sum_{w \in V} P^k(u,w)P^k(w,v) = \sum_{w \in V} P^k(u,w)P^k(v,w), \quad (1)$$

where the last equality follows from the fact that G is an undirected regular graph.

Since G is vertex-transitive, the two vectors

$$(P^k(u,w))_{w \in V} \qquad \text{and} \qquad (P^k(v,w))_{w \in V}$$

can be obtained from each other by permuting the coordinates. Therefore, by the claim above, the maximum possible value of the sum in the right-hand side of (1) is when $u = v$, completing the proof of the theorem. ■

Part II

TOPICS

10

Random Graphs

Let n be a positive integer, $0 \le p \le 1$. The random graph $G(n,p)$ is a probability space over the set of graphs on the vertex set $\{1,\ldots,n\}$ determined by

$$\Pr[\{i,j\} \in G] = p,$$

with these events mutually independent. This model is often used in the probabilistic method for proving the existence of certain graphs. In this chapter, we study the properties of $G(n,p)$ for their own sake.

Random graphs is an active area of research that combines probability theory and graph theory. The subject began in 1960 with the monumental paper "On the Evolution of Random Graphs" by Paul Erdős and Alfred Rényi. The book *Random Graphs* by Béla Bollobás (1985) is the standard source for the field. In this chapter, we explore only a few of the many topics in this fascinating area.

There is a compelling dynamic model for random graphs. For all pairs i,j, let $x_{i,j}$ be selected uniformly from $[0,1]$, the choices mutually independent. Imagine p going from 0 to 1. Originally, all potential edges are "off." The edge from i to j (which we may imagine as a neon light) is turned on when p reaches $x_{i,j}$ and then stays on. At $p=1$, all edges are "on." At time p, the graph of all "on" edges has distribution $G(n,p)$. As p increases, $G(n,p)$ *evolves* from empty to full.

In their original paper, Erdős and Rényi let $G(n,e)$ be the random graph with n vertices and precisely e edges. Again there is a dynamic model: Begin with no edges and add edges randomly one by one until the graph becomes full. Generally, $G(n,e)$ will have very similar properties as $G(n,p)$ with $p \sim e/\binom{n}{2}$. We will work on the probability model exclusively.

1. SUBGRAPHS

The term *the random graph* is, strictly speaking, a misnomer. $G(n,p)$ is a probability space over graphs. Given any graph theoretic property A, there will be a probability that $G(n,p)$ satisfies A, which we write $\Pr[G(n,p) \models A]$. When A is monotone, $\Pr[G(n,p) \models A]$ is a monotone function of p. As an instructive example, let A be the event "G is triangle-free." Let X be the number of triangles contained in $G(n,p)$. Linearity of expectation gives

$$E[X] = \binom{n}{3} p^3.$$

This suggests the parametrization $p = c/n$. Then

$$\lim_{n\to\infty} E[X] = \lim_{n\to\infty} \binom{n}{3} p^3 = c^3/6.$$

It turns out that the distribution of X is asymptotically Poisson. In particular,

$$\lim_{n\to\infty} \Pr[G(n,p) \models A] = \lim_{n\to\infty} \Pr[X = 0] = e^{-c^3/6}.$$

Note that

$$\lim_{c\to 0} e^{-c^3/6} = 1,$$

$$\lim_{c\to\infty} e^{-c^3/6} = 0.$$

When $p = 10^{-6}/n$, $G(n,p)$ is very unlikely to have triangles and when $p = 10^6/n$, $G(n,p)$ is very likely to have triangles. In the dynamic view, the first triangles almost always appear at $p = \Theta(1/n)$. If we take a function such as $p(n) = n^{-.9}$ with $p(n) \gg n^{-1}$, then $G(n,p)$ will almost always have triangles. Occasionally, we will abuse notation and say, for example, that $G(n, n^{-.9})$ contains a triangle—this meaning that the probability that it contains a triangle approaches 1 as n approaches infinity. Similarly, when $p(n) \ll n^{-1}$, for instance, $p(n) = 1/(n \ln n)$, then $G(n,p)$ will almost always not contain a triangle and we abuse notation and say that $G(n, 1/(n \ln n))$ is triangle-free. It was a central observation of Erdős and Rényi (1960) that many natural graph theoretic properties become true in a very narrow range of p. They made the following key definition.

Definition. $r(n)$ is called a threshold function for a graph theoretic property A if

(i) When $p(n) \ll r(n)$, $\lim_{n\to\infty} \Pr[G(n,p) \models A] = 0$,
(ii) When $p(n) \gg r(n)$, $\lim_{n\to\infty} \Pr[G(n,p) \models A] = 1$,

or visa versa.

In our example, $1/n$ is a threshold function for A. Note that the threshold function, when one exists, is not unique. We could equally have said that $10/n$ is a threshold function for A.

Let's approach the problem of $G(n, c/n)$ being triangle-free once more. For every set S of three vertices, let B_S be the event that S is a triangle. Then $\Pr[B_S] = p^3$. Then "triangle-freeness" is precisely the conjunction $\bigwedge \overline{B}_S$ over all S. If the B_S were mutually independent, then we *would* have

$$\Pr\left[\bigwedge \overline{B}_S\right] = \prod \left[\overline{B}_S\right] = (1-p^3)^{\binom{n}{3}} \sim e^{-\binom{n}{3}p^3} \to e^{-c^3/6}.$$

The reality is that the B_S are not mutually independent, though when $|S \cap T| \leq 1$, B_S and B_T are mutually independent.

We apply Janson's inequality, Theorem 1.1 of Chapter 8. In the notation of Section 1 of that chapter, $I = \{S \subset V(G) : |S| = 3\}$ and $S \sim T$ if and only if $|S \cap T| = 2$. Here $\epsilon = p^3 = o(1)$, $\mu = \binom{n}{3}p^3 \sim c^3/6$, and $M = \exp(-\mu(1+o(1))) = \exp(-c^3/6 + o(1))$. There are $6\binom{n}{4} = O(n^4)$ pairs S, T of triples with $S \sim T$. For each, $\Pr[B_S \wedge B_T] = p^5$. Thus

$$\Delta = O(n^4)p^5 = n^{-1+o(1)} = o(1).$$

When $\Delta = o(1)$ Janson's inequality sandwiches an asymptotic bound,

$$\lim_{n\to\infty} \Pr\left[\bigwedge \overline{B}_S\right] = \lim_{n\to\infty} M = e^{-c^3/6}.$$

Can we duplicate this success with the property A that G contains no (not necessarily induced) copy of a general given graph H? We use the definitions of balanced and strictly balanced of Chapter 4, Section 4.

Theorem 1.1. *Let H be a strictly balanced graph with v vertices, e edges, and a automorphisms. Let $c > 0$ be arbitrary. Let A be the property that G contains no copy of H. Then with $p = cn^{-v/e}$,*

$$\lim_{n\to\infty} \Pr[G(n,p) \models A] = e^{-c^e/a}.$$

Proof. Let A_α, $1 \leq \alpha \leq \binom{n}{v}v!/a$, range over the edge sets of possible copies of H and let B_α be the event $G(n,p) \supseteq A_\alpha$. We apply Janson's inequality. As

$$\lim_{n\to\infty} \mu = \lim_{n\to\infty} \binom{n}{v} \frac{v!\,p^e}{a} = \frac{c^e}{a},$$

we find

$$\lim_{n\to\infty} M = \exp[-c^e/a].$$

Now we examine (as in Chapter 4, Theorem 4.2)

$$\Delta = \sum_{\alpha\sim\beta} \Pr[B_\alpha \wedge B_\beta].$$

We split the sum according to the number of *vertices* in the intersection of copies α and β. Suppose they intersect in j vertices. If $j = 0$ or $j = 1$, then $A_\alpha \cap A_\beta = \emptyset$ so that $\alpha \sim \beta$ cannot occur. For $2 \leq j \leq v$, let f_j be the maximal $|A_\alpha \cap A_\beta|$ where $\alpha \sim \beta$ and α, β intersect in j vertices. As $\alpha \neq \beta$, $f_v < e$. When

$2 \le j \le v-1$, the critical observation is that $A_\alpha \cap A_\beta$ is a subgraph of H and hence, as H is strictly balanced,

$$\frac{f_j}{j} < \frac{e}{v}.$$

There are $O(n^{2v-j})$ choices of α, β intersecting in j points, since α, β are determined, except for order, by $2v-j$ points. For each such α, β,

$$\Pr[B_\alpha \wedge B_\beta] = p^{|A_\alpha \cup A_\beta|} = p^{2e-|A_\alpha \cap A_\beta|} \le p^{2e-f_j}.$$

Thus

$$\Delta = \sum_{j=2}^{v} O(n^{2v-j}) O\left(n^{-(v/e)(2e-f_j)}\right).$$

But

$$2v - j - \frac{v}{e}(2e - f_j) = \frac{vf_j}{e} - j < 0,$$

so each term is $o(1)$ and hence $\Delta = o(1)$. By Janson's inequality,

$$\lim_{n\to\infty} \Pr\left[\bigwedge \overline{B}_\alpha\right] = \lim_{n\to\infty} M = e^{-c^e/a},$$

completing the proof. ■

2. CLIQUE NUMBER

In this section, we fix $p = 1/2$ (other values yield similar results), and consider the clique number $\omega(G(n,p))$. For a fixed $c > 0$, let $n, k \to \infty$ so that

$$\binom{n}{k} 2^{-\binom{k}{2}} \to c.$$

As a first approximation,

$$n \sim \frac{k}{e\sqrt{2}} \sqrt{2}^k$$

and

$$k \sim \frac{2 \ln n}{\ln 2}.$$

Here $\mu \to c$, so $M \to e^{-c}$. The Δ term was examined in Chapter 4, Section 5. For this k, $\Delta = o(E[X]^2)$ and so $\Delta = o(1)$. Therefore

$$\lim_{n,k\to\infty} \Pr[\omega(G(n,p)) < k] = e^{-c}.$$

Being more careful, let $n_0(k)$ be the minimum n for which

$$\binom{n}{k} 2^{-\binom{k}{2}} \ge 1.$$

Observe that for this n, the left-hand side is $1+o(1)$. Note that $\binom{n}{k}$ grows, in n, like n^k. For any $\lambda \in (-\infty, +\infty)$, if

$$n = n_0(k)\left[1+\frac{\lambda+o(1)}{k}\right],$$

then

$$\binom{n}{k}2^{-\binom{k}{2}} = \left[1+\frac{\lambda+o(1)}{k}\right]^k = e^{\lambda}+o(1),$$

and so

$$\Pr[\omega(G(n,p))<k] = e^{-e^{\lambda}}+o(1).$$

As λ ranges from $-\infty$ to $+\infty$, $e^{-e^{\lambda}}$ ranges from 1 to 0. As $n_0(k+1)\sim\sqrt{2}n_0(k)$, the ranges will not "overlap" for different k. More precisely, let K be arbitrarily large and set

$$I_k = \left[n_0(k)\left[1-\frac{K}{k}\right], n_0(k)\left[1+\frac{K}{k}\right]\right].$$

For $k \geq k_0(K)$, $I_{k-1}\cap I_k = \emptyset$. Suppose $n \geq n_0(k_0(K))$. If n lies between the intervals (which occurs for "most" n), which we denote by $I_k < n < I_{k+1}$, then

$$\Pr[\omega(G(n,p))<k] \leq e^{-e^{K}}+o(1),$$

nearly 0, and

$$\Pr[\omega(G(n,p))<k+1] \geq e^{-e^{-K}}+o(1),$$

nearly 1, so that

$$\Pr[\omega(G(n,p))=k] \geq e^{-e^{-K}}-e^{-e^{K}}+o(1),$$

nearly 1. When $n \in I_k$, we still have $I_{k-1} < n < I_{k+1}$, so that

$$\Pr[\omega(G(n,p))=k \text{ or } k-1] \geq e^{-e^{-K}}-e^{-e^{K}}+o(1),$$

nearly 1. As K may be made arbitrarily large, this yields the celebrated 2-point concentration theorem on clique number, Corollary 5.2 in Chapter 4, Section 5. Note, however, that for most n, the concentration of $\omega(G(n,1/2))$ is actually on a single value!

3. CHROMATIC NUMBER

In this section, we fix $p = 1/2$ (there are similar results for other p) and let G be the random graph $G(n,1/2)$. We shall find bounds on the chromatic number $\chi(G)$. A different derivation of the main result of this section is presented in Chapter 7, Section 3. Set

$$f(k) = \binom{n}{k}2^{-\binom{k}{2}}.$$

Let $k_0 = k_0(n)$ be that value for which

$$f(k_0 - 1) > 1 > f(k_0).$$

Then $n = \sqrt{2}^{k(1+o(1))}$, so for $k \sim k_0$,

$$\frac{f(k+1)}{f(k)} = \frac{n}{k} 2^{-k}(1+o(1)) = n^{-1+o(1)}.$$

Set

$$k = k(n) = k_0(n) - 4$$

so that

$$f(k) > n^{3+o(1)}.$$

Now we use the generalized Janson inequality (Theorem 1.2 of Chapter 8) to estimate $\Pr[\omega(G) < k]$. Here $\mu = f(k)$. (Note that Janson's inequality gives a lower bound of $2^{-f(k)} = 2^{-n^{3+o(1)}}$ to this probability, but this is way off the mark, since with probability $2^{-\binom{n}{2}}$ the random G is empty!) The value Δ was examined in Chapter 4, Section 5, where

$$\frac{\Delta}{\mu^2} = \frac{\Delta^*}{\mu} = \sum_{i=2}^{k-1} g(i).$$

There $g(2) \sim k^4/n^2$ and $g(k-1) \sim 2kn2^{-k}/\mu$ were the dominating terms. In our instance, $\mu > n^{3+o(1)}$ and $2^{-k} = n^{-2+o(1)}$, so $g(2)$ dominates and

$$\Delta \sim \frac{\mu^2 k^4}{n^2}.$$

Hence we bound the *clique* number probability

$$\Pr[\omega(G) < k] < \exp\left(\frac{-\mu^2(1+o(1))}{2\Delta}\right) = \exp\left(-n^{2+o(1)}\right)$$

as $k = \Theta(\ln n)$. (The possibility that G is empty gives a lower bound, so that we may say the probability is $\exp(-n^{2+o(1)})$, though a $o(1)$ in the hyperexponent leaves lots of room.)

Theorem 3.1 (Bollobás [1988]). *Almost always,*

$$\chi(G) \sim \frac{n}{2\log_2 n}.$$

Proof. Let $\alpha(G) = \omega(\overline{G})$ denote, as usual, the independence number of G. The complement of G has the same distribution $G(n, 1/2)$. Hence $\alpha(G) \le (2 + o(1))\log_2 n$ almost always. Thus

$$\chi(G) \ge \frac{n}{\alpha(G)} \ge \frac{n}{2\log_2 n}(1 + o(1))$$

almost always.

The reverse inequality was an open question for a full quarter-century! Set $m = \lfloor n/\ln^2 n \rfloor$. For any set S of m vertices, the restriction $G|_S$ has the distribution of $G(m,1/2)$. Let $k = k(m) = k_0(m) - 4$ as above. Note

$$k \sim 2\log_2 m \sim 2\log_2 n.$$

Then

$$\Pr[\alpha[G|_S] < k] < \exp(-m^{2+o(1)}).$$

There are $\binom{n}{m} < 2^n = 2^{m^{1+o(1)}}$ such sets S. Hence

$$\Pr[\alpha[G|_S] < k \text{ for some } m\text{-set } S] < 2^{m^{1+o(1)}} \exp(-m^{2+o(1)}) = o(1).$$

That is, almost always *every* set of m vertices contains a k-element independent set.

Now suppose G has this property. We pull out k-element independent sets and give each a distinct color until there are less than m vertices left. Then we give each point a distinct color. By this procedure,

$$\begin{aligned}\chi(G) &\le \left\lceil \frac{n-m}{k} \right\rceil + m \le \frac{n}{k} + m \\ &= \frac{n}{2\log_2 n}(1+o(1)) + o\left(\frac{n}{\log_2 n}\right) \\ &= \frac{n}{2\log_2 n}(1+o(1)),\end{aligned}$$

and this occurs for almost all G. ■

4. BRANCHING PROCESSES

Paul Erdős and Alfred Rényi, in their original 1960 paper, discovered that the random graph $G(n,p)$ undergoes a remarkable change at $p = 1/n$. Speaking roughly, let first $p = c/n$ with $c < 1$. Then $G(n,p)$ will consist of small components, the largest of which is of size $\Theta(\ln n)$. But now suppose $p = c/n$ with $c > 1$. In that short amount of "time" many of the components will have joined together to form a "giant component" of size $\Theta(n)$. The remaining vertices are still in small components, the largest of which has size $\Theta(\ln n)$. They dubbed this phenomenon the *double jump*. We prefer the descriptive term *phase transition* because of the connections to percolation (e.g., freezing) in mathematical physics.

To better understand the phase transition, we make a lengthy detour into the subject of branching processes. Imagine that we are in a unisexual universe and we start with a single organism. Imagine that this organism has a number of children given by a given random variable Z. (For us, Z will be Poisson with mean c.) These children then themselves have children, the number again being determined by Z. These grandchildren then have children, etc. As $Z = 0$ will have nonzero probability, there will be some chance that the line dies out entirely. We want to study the total number of organisms in this process, with particular eye to whether or not the process continues forever. (The original application of this model was to a study of the—gasp!—male line of British peerage.)

Now let's be more precise. Let $Z_1, Z_2, \ldots$ be independent random variables, each with distribution Z. Define $Y_0, Y_1, \ldots$ by the recursion

$$Y_0 = 1,$$
$$Y_i = Y_{i-1} + Z_i - 1,$$

and let T be the least t for which $Y_t = 0$. If no such t exists (the line continuing forever), we say $T = +\infty$. The Y_i and Z_i mirror the branching process as follows. We view all organisms as living or dead. Initially, there is one live organism and no dead ones. At each time unit, we select one of the live organisms, it has Z_i children, and then it dies. The number Y_i of live organisms at time i is then given by the recursion. The process stops when $Y_t = 0$ (extinction), but it is a convenient fiction to define the recursion for all t. Note that T is not affected by this fiction, since once $Y_t = 0$, T has been defined. T (whether finite or infinite) is the total number of organisms, including the original, in this process. (A natural approach, found in many probability texts, is to have all organisms of a given generation have their children at once and study the number of children of each generation. While we may think of the organisms giving birth by generation, it will not affect our model.)

A major result of branching processes is that when $E[Z] = c < 1$ with probability 1, the process dies out ($T < \infty$), but when $E[Z] = c > 1$, then there is a nonzero probability that the process goes on forever ($T = \infty$). Intuitively, this makes sense: the values Y_i of the population form a Markov chain. When $c < 1$, there is drift "to the left," so eventually $Y_i = 0$, whereas when $c > 1$, there is drift "to the right," $Y_i \to +\infty$, and with finite probability the population never hits 0. If the process doesn't die early, the population is likely to just keep growing. We give a proof based on the large deviation bounds (Theorem A.15). Observe that $Z_1 + \cdots + Z_t$ has a Poisson distribution with mean ct. First suppose $c < 1$. For any t,

$$\Pr[T > t] \le \Pr[Y_t > 0] = \Pr[Z_1 + \cdots + Z_t \ge t] \le (1 - \delta)^t,$$

with $\delta > 0$. As $\lim_{t\to\infty} \Pr[T > t] = 0$, $\Pr[T = \infty] = 0$. Now suppose $c > 1$. Again using Theorem A.15,

$$\Pr[Y_t \le 0] = \Pr[Z_1 + \cdots + Z_t \le t] \le (1 - \delta)^t,$$

with a different $\delta > 0$. As $\sum_{t=1}^{\infty} (1 - \delta)^t$ converges, there is a t_0 with

$$\sum_{t=t_0}^{\infty} \Pr[Y_t \le 0] < 1.$$

Then

$$\sum_{t=0}^{\infty} \Pr[Y_t + t_0 - 1 \leq 0] < 1,$$

since for $t < t_0$, $Y_t + t_0 - 1 \geq 1 - t + t_0 - 1 > 0$ always. We now condition on the first organism having t_0 children. The conditional distribution of

$$Y_t = t_0 + (Z_2 - 1) + \cdots + (Z_t - 1)$$

is the unconditional distribution of

$$Y_{t-1} + (t_0 - 1) = t_0 + (Z_1 - 1) + \cdots + (Z_{t-1} - 1),$$

and so

$$\sum_{t=0}^{\infty} \Pr[Y_t \leq 0 \mid Z_1 = t_0] < 1.$$

With positive probability $Z_1 = t_0$, and then with positive probability all $Y_t > 0$, so that $\Pr[T = \infty] > 0$.

Generating functions give a more precise result. Let

$$p_i = \Pr[Z_j = i]$$

and define the generating function

$$p(x) = \sum_{i=0}^{\infty} p_i x^i.$$

In our case,

$$p_i = \frac{e^{-c} c^i}{i!},$$

so that

$$p(x) = \sum_{i=0}^{\infty} e^{-c} c^i x^i / i! = e^{c(x-1)}.$$

Let

$$q_i = \Pr[T = i]$$

and set

$$q(x) = \sum_{i=0}^{\infty} q_i x^i$$

(the sum not including the $i = \infty$ case). Conditioning on the first organism having s children, the generating function for the total number of offspring is $x(q(x))^s$. Hence

$$q(x) = \sum_{s=0}^{\infty} p_s x q(x)^s = x p[q(x)].$$

That is, the generating function $y = q(x)/x$ satisfies the functional equality $y = p(xy)$, i.e.,

$$y = e^{c(xy-1)}.$$

The extinction probability

$$y = \Pr[T < \infty] = \sum_{i=0}^{\infty} q_i = q(1) = \frac{q(1)}{1}$$

must satisfy

$$y = e^{c(y-1)} \tag{**}$$

For $c < 1$, this has the unique solution $y = 1$, corresponding to the certain extinction. For $c > 1$, there are two solutions, $y = 1$ and some $y \in (0,1)$. For $c > 1$, we let $f(c)$ denote that y satisfying (**), $0 < y < 1$. As $\Pr[T < \infty] < 1$, we know

$$\Pr[T < \infty] = f(c).$$

When a branching process dies, we call $H = (Z_1, \ldots, Z_T)$ the *history* of the process. A sequence $(z_1, \ldots, z_t)$ is a possible history if and only if the sequence y_i given by $y_0 = 1, y_i = y_{i-1} + z_i - 1$ has $y_i > 0$ for $0 \le i < t$ and $y_t = 0$. When Z is Poisson with mean λ,

$$\Pr[H = (z_1, \ldots, z_t)] = \prod_{i-1}^{t} \frac{e^{-\lambda}\lambda^{z_i}}{z_i!} = \frac{e^{-\lambda}(\lambda e^{-\lambda})^{t-1}}{\prod_{i=1}^{t} z_i!},$$

since $z_1 + \cdots + z_t = t - 1$.

We call $d < 1 < c$ a conjugate pair if

$$de^{-d} = ce^{-c}.$$

The function $f(x) = xe^{-x}$ increases from 0 to e^{-1} in $[0,1)$ and decreases back to 0 in $(1, \infty)$, so that all $c \neq 1$ have a unique conjugate. Let $c > 1$ and $y = \Pr[T < \infty]$ so that $y = e^{c(y-1)}$. Then $(cy)e^{-cy} = ce^{-c}$, so

$$d = cy.$$

Duality Principle. *Let $d < 1 < c$ be conjugates. The branching process with mean c, conditional on extinction, has the same distribution as the branching process with mean d.*

Proof. It suffices to show that for every history $H = (z_1, \ldots, z_t)$,

$$\frac{e^{-c}(ce^{-c})^{t-1}}{y \prod_{i=1}^{t} z_i!} = \frac{e^{-d}(de^{-d})^{t-1}}{\prod_{i=1}^{t} z_i!}.$$

This is immediate, as $ce^{-c} = de^{-d}$ and $ye^{-d} = ye^{-cy} = e^{-c}$. ■

5. THE GIANT COMPONENT

Now let's return to random graphs. We define a procedure to find the component $C(v)$ containing a given vertex v in a given graph G. We are motivated by Karp (1990), in which this approach is applied to random digraphs. In this procedure, vertices will be live, dead, or neutral. Originally, v is live and all other vertices are neutral, time $t = 0$, and $Y_0 = 1$. Each time unit t we take a live vertex w and check all pairs $\{w, w'\}$, w' neutral, for membership in G. If $\{w, w'\} \in G$, we make w' live, otherwise it stays neutral. After searching all neutral w', we set w dead and let Y_t equal the new number of live vertices. When there are no live vertices, the process terminates and $C(v)$ is the set of dead vertices. Let Z_t be the number of w' with $\{w, w'\} \in G$ so that

$$Y_0 = 1,$$
$$Y_t = Y_{t-1} + Z_t - 1.$$

With $G = G(n, p)$, each neutral w' has independent probability p of becoming live. Here, critically, no pair $\{w, w'\}$ is ever examined twice, so that the conditional probability for $\{w, w'\} \in G$ is always p. As $t-1$ vertices are dead and Y_{t-1} are live,

$$Z_t \sim B[n - (t-1) - Y_{t-1}, p].$$

Let T be the least t for which $Y_t = 0$. Then $T = |C(v)|$. As in Section 4, we continue the recursive definition of Y_t, this time for $0 \le t \le n$.

Claim 5.1. *For all t,*

$$Y_t \sim B[n-1, 1-(1-p)^t] + 1 - t.$$

Proof. It is more convenient to deal with

$$N_t = n - t - Y_t,$$

the number of neutral vertices at time t, and show, equivalently,

$$N_t \sim B[n-1, (1-p)^t].$$

This is reasonable, since each $w \neq v$ has independent probability $(1-p)^t$ of staying neutral t times. Formally, as $N_0 = n-1$ and

$$\begin{aligned} N_t = n - t - Y_t &= n - t - B[n-(t-1) - Y_{t-1}, p] - Y_{t-1} + 1 \\ &= N_{t-1} - B[N_{t-1}, p] \\ &= B[N_{t-1}, 1-p], \end{aligned}$$

the result follows by induction. ∎

We set $p = c/n$. When t and Y_{t-1} are small, we may approximate Z_t by $B[n,c/n]$, which is approximately Poisson with mean c. Basically, small components will have size distribution as in the branching process of Section 4. The analogy must break down for $c > 1$, as the branching process may have an infinite population, whereas $|C(v)|$ is surely at most n. Essentially, those v for which the branching process for $C(v)$ does not "die early" all join together to form the giant component.

Fix c. Let $Y_0^*, Y_1^*, \ldots, T^*, Z_1^*, Z_2^*, \ldots, H^*$ refer to the branching process of Section 4 and let $Y_0, Y_1, \ldots, T, Z_1, Z_2, \ldots, H$ refer to the random graph process. For any possible history $(z_1, \ldots, z_t)$,

$$\Pr[H^* = (z_1, \ldots, z_t)] = \prod_{i=1}^{t} \Pr[Z^* = z_i],$$

where Z^* is Poisson with mean c, while

$$\Pr[H = (z_1, \ldots, z_t)] = \prod_{i=1}^{t} \Pr[Z_i = z_i],$$

where Z_i has binomial distribution $B[n - 1 - z_1 - \cdots - z_{i-1}, c/n]$. The Poisson distribution is the limiting distribution of binomials. When $m = m(n) \sim n$ and c, i are fixed,

$$\lim_{n\to\infty} \Pr[B[m,c/n] = i] = \lim_{n\to\infty} \binom{m}{z} \left(\frac{c}{n}\right)^z \left(1 - \frac{c}{n}\right)^{m-z} = e^{-c}c^z/z!.$$

Hence

$$\lim_{n\to\infty} \Pr[H = (z_1, \ldots, z_t)] = \Pr[H^* = (z_1, \ldots, z_t)].$$

Assume $c < 1$. For any fixed t, $\lim_{n\to\infty} \Pr[T = t] = \Pr[T^* = t]$. We now bound the size of the largest component. For any t,

$$\Pr[T > t] \leq \Pr[Y_t > 0] = \Pr[B[n-1, 1-(1-p)^t] \geq t] \leq \Pr[B[n, tc/n] \geq t],$$

as $1 - (1-p)^t \leq tp$ and $n - 1 < n$. By large deviation result (Theorem A.14),

$$\Pr[T > t] < e^{-\alpha t},$$

where $\alpha = \alpha(c) > 0$. Let $\beta = \beta(c)$ satisfy $\alpha\beta > 1$. Then

$$\Pr[T > \beta \ln n] < n^{-\alpha\beta} = o(n^{-1}).$$

There are n choices for initial vertex v. Thus almost always, *all* components have size $O(\ln n)$.

Now assume $c > 1$. For any fixed t, $\lim_{n\to\infty} \Pr[T = t] = \Pr[T^* = t]$, but what corresponds to $T^* = \infty$? For $t = o(n)$, we may estimate $1 - (1-p)^t \sim pt$ and $n - 1 \sim n$ so that

$$\Pr[Y_t \leq 0] = \Pr[B[n-1, 1-(1-p)^t] \leq t-1] \sim \Pr[B[n, tc/n] \leq t]$$

drops exponentially in t by large deviation results. When $t = \alpha n$, we estimate $1-(1-p)^t$ by $1-e^{-c\alpha}$. The equation $1-e^{-c\alpha} = \alpha$ has solution $\alpha = 1-y$, where y is the extinction probability of Section 4. For $\alpha < 1-y$, $1-e^{-c\alpha} > \alpha$, and

$$\Pr[Y_t \leq 0] \sim \Pr[B[n, 1-e^{-c\alpha}] \leq \alpha n]$$

is exponentially small, while for $\alpha > 1-y$, $1-e^{-c\alpha} < \alpha$ and $\Pr[Y_t \leq 0] \sim 1$. Thus almost always, $Y_t = 0$ for some $t \sim (1-y)n$. Basically, $T^* = \infty$ corresponds to $T \sim (1-y)n$. Let $\epsilon, \delta > 0$ be arbitrarily small. With somewhat more care to the bounds, we may show that there exists t_0 so that for n sufficiently large,

$$\Pr[t_0 < T < (1-\delta)n(1-y) \text{ or } T > (1+\delta)n(1-y)] < \epsilon.$$

Pick t_0 sufficiently large so that

$$y - \epsilon \leq \Pr[T^* \leq t_0] \leq y.$$

Then as $\lim_{n\to\infty}\Pr[T \leq t_0] = \Pr[T^* \leq 0]$ for n sufficiently large,

$$y - 2\epsilon \leq \Pr[T \leq t_0] \leq y + \epsilon$$

$$1 - y - 2\epsilon \leq \Pr[(1-\delta)n(1-y) < T < (1+\delta)n(1-y)] < 1 - y + 3\epsilon.$$

Now we expand our procedure to find graph components. We start with $G \sim G(n,p)$, select $v = v_1 \in G$, and compute $C(v_1)$ as before. Then we delete $C(v_1)$, pick $v_2 \in G - C(v_1)$, and iterate. At each stage, the remaining graph has distribution $G(m,p)$, where m is the number of vertices. (Note, critically, that no pairs $\{w, w'\}$ in the remaining graph have been examined and so it retains its distribution.) Call a component $C(v)$ small if $|C(v)| \leq t_0$, giant if $(1-\delta)n(1-y) < |C(v)| < (1+\delta)n(1-y)$, and otherwise failure. Pick $s = s(\epsilon)$ with $(y+\epsilon)^s < \epsilon$. (For ϵ small, $s \sim K \ln \epsilon^{-1}$.) Begin this procedure with the full graph and terminate it when either a giant component or a failure component is found or when s small components are found. At each stage, as only small components have thus far been found, the number of remaining points is $m = n - O(1) \sim n$, so the conditional probabilities of small, giant, and failure remain asymptotically the same. The chance of ever hitting a failure component is therefore $\leq s\epsilon$ and the chance of hitting all small components is $\leq (y+\epsilon)^s \leq \epsilon$, so that with probability at least $1-\epsilon'$, where $\epsilon' = (s+1)\epsilon$ may be made arbitrarily small, we find a series of less than s small components followed by a giant component. The remaining graph has $m \sim yn$ points and $pm \sim cy = d$, the conjugate of c as defined in Section 4. As $d < 1$, the previous analysis gives the maximal components. In summary: Almost always, $G(n, c/n)$ has a giant component of size $\sim (1-y)n$ and all other components of size $O(\ln n)$. Furthermore, the duality principle of Section 4 has a discrete analog.

Discrete Duality Principle. *Let $d < 1 < c$ be conjugates. The structure of $G(n,c/n)$ with its giant component removed is basically that of $G(m,d/m)$ where m, the number of vertices not in the giant component, satisfies $m \sim ny$.*

The small components of $G(n,c/n)$ can also be examined from a static view. For a fixed k, let X be the number of tree components of size k. Then

$$E[X] = \binom{n}{k} k^{k-2} \left(\frac{c}{n}\right)^{k-1} \left(1 - \frac{c}{n}\right)^{k(n-k)+\binom{k}{2}-(k-1)}.$$

Here we use the nontrivial fact, due to Cayley, that there are k^{k-2} possible trees on a given k-set. For c,k fixed,

$$E[X] \sim n \frac{e^{-ck} k^{k-2} c^{k-1}}{k!}.$$

As trees are strictly balanced, a second moment method gives $X \sim E[X]$ almost always. Thus $\sim p_k n$ points lie in tree components of size k, where

$$p_k = \frac{e^{-ck}(ck)^{k-1}}{k!}.$$

It can be shown analytically that $p_k = \Pr[T = k]$ in the branching process with mean c of Section 4. Let Y_k denote the number of cycles of size k and Y the total number of cycles. Then

$$E[Y_k] = \frac{(n)_k}{2k} \left(\frac{c}{n}\right)^k \sim \frac{c^k}{2k}$$

for fixed k. For $c < 1$,

$$E[Y] = \sum E[Y_k] \to \sum_{k=1}^{\infty} \frac{c^k}{2k}$$

has a finite limit, whereas for $c > 1$, $E[Y] \to \infty$. Even for $c > 1$, for any fixed k, the number of k-cycles has a limiting expectation and so do not asymptotically affect the number of components of a given size.

6. INSIDE THE PHASE TRANSITION

In the evolution of the random graph $G(n,p)$, a crucial change takes place in the vicinity of $p = c/n$ with $c = 1$. The small components at that time are rapidly joining together to form a giant component. This corresponds to the branching process when births are Poisson with mean 1. There the number T of organisms will be finite almost always and yet have infinite expectation. No wonder that the situation for random graphs is extremely delicate. In recent years, there has been much interest in looking "inside" the phase transition at the growth of the largest components. (See, e.g., Luczak [1990].) The appropriate parametrization is, perhaps surprisingly,

$$p = \frac{1}{n} + \frac{\lambda}{n^{4/3}}.$$

When $\lambda = \lambda(n) \to -\infty$, the phase transition has not yet started. The largest components are $o(n^{2/3})$ and there are many components of nearly the largest size. When $\lambda = \lambda(n) \to +\infty$, the phase transition is over—a largest component, of size $\gg n^{2/3}$, has emerged and all other components are of size $o(n^{2/3})$. Let's fix λ and c and let X be the number of tree components of size $k = cn^{2/3}$. Then

$$E[X] = \binom{n}{k} k^{k-2} \left(\frac{c}{n}\right)^{k-1} \left(1 - \frac{c}{n}\right)^{k(n-k)+\binom{k}{2}-(k-1)}.$$

Watch the terms cancel!

$$\binom{n}{k} = \frac{(n)_k}{k!} \sim \frac{n^k e^k}{k^k\sqrt{2\pi k}} \prod_{i=1}^{k-1} \left(1 - \frac{i}{n}\right).$$

For $i < k$,

$$-\ln\left(1 - \frac{i}{n}\right) = \frac{i}{n} + \frac{i^2}{2n^2} + O\left(\frac{i^3}{n^3}\right),$$

so that

$$\sum_{i=1}^{k-1} -\ln\left(1 - \frac{i}{n}\right) = \frac{k^2}{2n} + \frac{k^3}{6n^2} + o(1) = \frac{k^2}{2n} + \frac{c^3}{6} + o(1).$$

Also

$$p^{k-1} = n^{1-k}\left(1 + \frac{\lambda}{n^{1/3}}\right)^{k-1},$$

$$(k-1)\ln\left(1 + \frac{\lambda}{n^{1/3}}\right) = (k-1)\left(\frac{\lambda}{n^{1/3}} - \frac{\lambda^2}{2n^{2/3}} + O(n^{-1})\right)$$

$$= \frac{\lambda k}{n^{1/3}} - \frac{\lambda^2 c}{2} + o(1).$$

Also

$$\ln(1-p) = -p + O(n^{-2}) = -\frac{1}{n} - \frac{\lambda}{n^{4/3}} + O(n^{-2})$$

and

$$k(n-k) + \binom{k}{2} - (k-1) = kn - \frac{k^2}{2} + O(n^{2/3}),$$

so that

$$\left[k(n-k) + \binom{k}{2} - (k-1)\right]\ln(1-p) = -k + \frac{k^2}{2n} - \frac{\lambda k}{n^{1/3}} + \frac{\lambda c^2}{2} + o(1)$$

and

$$E[X] \sim \frac{n^k k^{k-2}}{k^k \sqrt{2\pi k}\, n^{k-1}} e^A,$$

where

$$\begin{aligned} A &= k - \frac{k^2}{2n} - \frac{c^3}{6} + \frac{\lambda k}{n^{1/3}} - \frac{\lambda^2 c}{2} - k + \frac{k^2}{2n} - \frac{\lambda k}{n^{1/3}} + \frac{\lambda c^2}{2} + o(1) \\ &= -\frac{c^3}{6} - \frac{\lambda^2 c}{2} + \frac{\lambda c^2}{2} + o(1), \end{aligned}$$

so that

$$E[X] \sim n^{-2/3} \exp\left(-\frac{c^3}{6} - \frac{\lambda^2 c}{2} + \frac{\lambda c^2}{2}\right) c^{-5/2} (2\pi)^{-1/2}$$

For any particular such k, $E[X] \to 0$, but if we sum k between $cn^{2/3}$ and $(c+dc)n^{2/3}$, we multiply by $n^{2/3}dc$. Going to the limit gives an integral: For any fixed a, b, λ, let X be the number of tree components of size between $an^{2/3}$ and $bn^{2/3}$. Then

$$\lim_{n\to\infty} E[X] = \int_a^b \exp\left(-\frac{c^3}{6} - \frac{\lambda^2 c}{2} + \frac{\lambda c^2}{2}\right) c^{-5/2} (2\pi)^{-1/2} dc.$$

The large components are not all trees. E. M. Wright (1977) proved that for fixed l, there are asymptotically $c_l k^{k-2+(3/2)l}$ connected graphs on k points with $k-1+l$ edges, where c_l was given by a specific recurrence. Asymptotically in l, $c_l = l^{-l/2(1+o(1))}$. The calculation for $X^{(l)}$, the number of such components on k vertices, leads to extra factors of $c_l k^{(3/2)l}$ and n^{-l}, which gives $c_l c^{(3/2)l}$. For fixed a, b, λ, l, the number $X^{(l)}$ of components of size between $an^{2/3}$ and $bn^{2/3}$ with $l-1$ more edges than vertices satisfies

$$\lim_{n\to\infty} E[X^{(l)}] = \int_a^b \exp\left(-\frac{c^3}{6} - \frac{\lambda^2 c}{2} + \frac{\lambda c^2}{2}\right) c^{-5/2} (2\pi)^{-1/2} (c_l c^{(3/2)l})\, dc,$$

and, letting X^* be the total number of components of size between $an^{2/3}$ and $bn^{2/3}$,

$$\lim_{n\to\infty} E[X^*] = \int_a^b \exp\left(-\frac{c^3}{6} - \frac{\lambda^2 c}{2} + \frac{\lambda c^2}{2}\right) c^{-5/2} (2\pi)^{-1/2} g(c)\, dc,$$

where

$$g(c) = \sum_{l=0}^{\infty} c_l c^{(3/2)l},$$

a sum convergent for all c (here $c_0 = 1$). A component of size $\sim cn^{2/3}$ will have probability $c_l c^{(3/2)l}/g(c)$ of having $l-1$ more edges than vertices, independent of λ. As $\lim_{c\to 0} g(c) = 1$, most components of size $\epsilon n^{2/3}$, $\epsilon \ll 1$, are trees, but as c gets bigger, the distribution on l moves inexorably higher.

An Overview

For any fixed λ, the sizes of the largest components are of the form $cn^{2/3}$, with a distribution over the constant. For $\lambda = -10^6$, there is some positive limiting probability that the largest component is bigger than $10^6 n^{2/3}$, and for $\lambda = +10^6$, there is some positive limiting probability that the largest component is smaller than $10^{-6}n^{2/3}$, though both these probabilities are minuscule. The functions integrated have a pole at $c = 0$, reflecting the notion that for any λ, there should be many components of size near $\epsilon n^{2/3}$ for $\epsilon = \epsilon(\lambda)$ appropriately small. When λ is large negative (e.g., -10^6), the largest component is likely to be $\epsilon n^{2/3}$, ϵ small, and there will be many components of nearly that size. The nontree components will be a negligible fraction of the tree components.

Now consider the evolution of $G(n,p)$ in terms of λ. Suppose that at a given λ, there are components of size $c_1 n^{2/3}$ and $c_2 n^{2/3}$. When we move from λ to $\lambda + d\lambda$, there is a probability $c_1 c_2 d\lambda$ that they will merge. Components have a peculiar gravitation in which the probability of merging is proportional to their sizes. With probability $(c_1^2/2)d\lambda$, there will be a new internal edge in a component of size $c_1 n^{2/3}$, so that large components rarely remain trees. Simultaneously, big components are eating up other vertices.

With $\lambda = -10^6$, say, we have feudalism. Many small components (castles) are each vying to be the largest. As λ increases, the components increase in size and a few large components (nations) emerge. An already large France has much better chances of becoming larger than a smaller Andorra. The largest components tend strongly to merge and by $\lambda = +10^6$, it is very likely that a giant component, Roman Empire, has emerged. With high probability, this component is nevermore challenged for supremacy but continues absorbing smaller components until full connectivity—One World—is achieved.

7. ZERO–ONE LAWS

In this section, we restrict our attention to graph theoretic properties expressible in the *first-order theory* of graphs. The language of this theory consists of variables $(x, y, z, \ldots)$, which always represent vertices of a graph, equality and adjacency $(x = y, x \sim y)$, the usual Boolean connectives $(\wedge, \neg, \ldots)$ and universal and existential quantification $(\forall_x, \exists_y)$. Sentences must be finite. As examples, one can express the property of containing a triangle

$$\exists_x \exists_y \exists_z [x \sim y \wedge x \sim z \wedge y \sim z]$$

having no isolated point

$$\forall_x \exists_y [x \sim y]$$

and having radius at most 2

$$\exists_x \forall_y [\neg(y = x) \wedge \neg(y \sim x) \longrightarrow \exists_z [z \sim y \wedge y \sim x]].$$

For any property A and any n, p, we consider the probability that the random graph $G(n,p)$ satisfies A, denoted

$$\Pr[G(n,p) \models A].$$

Our objects in this section will be the theorem of Glebskii et al. (1969) and, independently, Fagin (1976) (Theorem 7.1), and that of Shelah and Spencer (1988) (Theorem 7.2).

Theorem 7.1. *For any fixed p, $0 < p < 1$, and any first-order A,*

$$\lim_{n\to\infty} \Pr[G(n,p) \models A] = 0 \text{ or } 1.$$

Theorem 7.2. *For any irrational α, $0 < \alpha < 1$, setting $p = p(n) = n^{-\alpha}$,*

$$\lim_{n\to\infty} \Pr[G(n,p) \models A] = 0 \text{ or } 1.$$

Both proofs are only outlined.

We shall say that a function $p = p(n)$ satisfies the zero–one law if the above equality holds for every first-order A.

The Glebskii–Fagin theorem has a natural interpretation when $p = .5$, as then $G(n,p)$ gives equal weight to every (labeled) graph. It then says that any first-order property A holds for either almost all graphs or for almost no graphs. The Shelah–Spencer theorem may be interpreted in terms of threshold functions. The general results of Section 1 give, as one example, that $p = n^{-2/3}$ is a threshold function for containment of a K_4. That is, when $p \ll n^{-2/3}$, $G(n,p)$ almost surely does not contain a K_4, whereas when $p \gg n^{-2/3}$, it almost surely does contain a K_4. In between, say at $p = n^{-2/3}$, the probability is between 0 and 1, in this case $1 - e^{-1/24}$. The (admittedly rough) notion is that *at* a threshold function, the zero–one law will not hold, and so to say that $p(n)$ satisfies the zero–one law is to say that $p(n)$ is not a threshold function—that it is a boring place in the evolution of the random graph, at least through the spectacles of the first-order language. In stark terms: What happens in the evolution of $G(n,p)$ at $p = n^{-\pi/7}$? The answer: Nothing!

Our approach to zero–one laws will be through a variant of the Ehrenfeucht game, which we now define. Let G, H be two vertex disjoint graphs and t a positive integer. We define a perfect information game, denoted EHR$[G,H,t]$, with two players, denoted Spoiler and Duplicator. The game has t rounds. Each round has two parts. First, Spoiler selects either a vertex $x \in V(G)$ or a vertex $y \in V(H)$. He chooses which graph to select the vertex from. Then Duplicator must select a vertex in the other graph. At the end of the t rounds, t vertices have been selected from each graph. Let $x_1, \ldots, x_t$ be the vertices selected from $V(G)$ and let $y_1, \ldots, y_t$ be the vertices selected from $V(H)$, where x_i, y_i are the vertices selected in the ith round. Then Duplicator wins if and only if the induced graphs on the selected vertices are order-isomorphic: that is, if for all $1 \le i < j \le t$,

$$\{x_i, x_j\} \in E(G) \longleftrightarrow \{y_i, y_j\} \in E(H).$$

As there are no hidden moves and no draws, one of the players must have a winning strategy and we will say that that player wins $\mathrm{EHR}[G,H,t]$.

Lemma 7.3. *For every first-order A, there is a $t = t(A)$ so that if G,H are any graphs with $G \models A$ and $H \models \neg A$, then Spoiler wins* $\mathrm{EHR}[G,H,t]$.

A detailed proof would require a formal analysis of the first-order language, so we give only an example. Let A be the property $\forall_x \exists_y [x \sim y]$ of not containing an isolated point and set $t = 2$. Spoiler begins by selecting an isolated point $y_1 \in V(H)$, which he can do as $H \models \neg A$. Duplicator must pick $x_1 \in V(G)$. As $G \models A$, x_1 is not isolated, so Spoiler may pick $x_2 \in V(G)$ with $x_1 \sim x_2$ and now Duplicator cannot pick a "duplicating" y_2.

Theorem 7.4. *A function $p = p(n)$ satisfies the zero–one law if and only if for every t, letting $G(n,p(n)), H(m,p(m))$ be independently chosen random graphs on disjoint vertex sets,*

$$\lim_{m,n \to \infty} \Pr[\text{Duplicator wins } \mathrm{EHR}[G(n,p(n)),H(m,p(m)),t]] = 1.$$

Remark. For any given choice of G,H, somebody must win $\mathrm{EHR}[G,H,t]$. (That is, there is no random play, the play is perfect.) Given this probability distribution over (G,H), there will be a probability that $\mathrm{EHR}[G,H,t]$ will be a win for Duplicator, and this must approach 1.

Proof. We prove only the "if" part. Suppose $p = p(n)$ did not satisfy the zero–one law. Let A satisfy

$$\lim_{n \to \infty} \Pr[G(n,p(n)) \models A] = c,$$

with $0 < c < 1$. Let $t = t(A)$ be as given by the lemma. With limiting probability $2c(1-c) > 0$, exactly one of $G(n,p(n))$, $H(n,p(n))$ would satisfy A and thus Spoiler would win, contradicting the assumption. This is not a full proof, since when the zero–one law is not satisfied, $\lim_{n\to\infty}\Pr[G(n,p(n)) \models A]$ might not exist. If there is a subsequence n_i on which the limit is $c \in (0,1)$, we may use the same argument. Otherwise, there will be two subsequences n_i, m_i on which the limit is 0 and 1, respectively. Then letting $n,m \to \infty$ through n_i, m_i, respectively, Spoiler will win $\mathrm{EHR}[G,H,t]$ with probability approaching 1. ■

Theorem 7.4 provides a bridge from logic to random graphs. To prove that $p = p(n)$ satisfies the zero–one law, we now no longer need to know anything about logic—we just have to find a good strategy for Duplicator.

We say that a graph G has the full level s extension property if for every distinct $u_1,\ldots,u_a,v_1,\ldots,v_b \in G$ with $a+b \leq s$, there is an $x \in V(G)$ with $\{x,u_i\} \in E(G)$, $1 \leq i \leq a$, and $\{x,v_j\} \notin V(G)$, $1 \leq j \leq b$. Suppose that G,H both have the full level $s-1$ extension property. Then Duplicator wins EHR$[G,H,s]$ by the following simple strategy. On the ith round, with $x_1,\ldots,$ $x_{i-1},y_1,\ldots,y_{i-1}$ already selected, and Spoiler picking, say, x_i, Duplicator simply picks y_i having the same adjacencies to the y_j, $j<i$ as x_i has to the x_j, $j<i$. The full extension property says that such a y_i will surely exist.

Theorem 7.5. *For any fixed* p, $0<p<1$, *and any* s, $G(n,p)$ *almost always has the full level* s *extension property.*

Proof. For every distinct $u_1,\ldots,u_a,v_1,\ldots,v_b,x \in G$ with $a+b \leq s$ let $E_{u_1,\ldots,u_a,v_1,\ldots,v_b,x}$ be the event that $\{x,u_i\} \in E(G)$, $1 \leq i \leq a$ and $\{x,v_j\} \notin V(G)$, $1 \leq j \leq b$. Then

$$\Pr[E_{u_1,\ldots,u_a,v_1,\ldots,v_b,x}] = p^a(1-p)^b.$$

Now define

$$E_{u_1,\ldots,u_a,v_1,\ldots,v_b} = \bigwedge_x \overline{E_{u_1,\ldots,u_a,v_1,\ldots,v_b,x}},$$

the conjunction over $x \neq u_1,\ldots,u_a,v_1,\ldots,v_b$. But these events are mutually independent over x, since they involve different edges. Thus

$$\Pr\left[\bigwedge_x \overline{E_{u_1,\ldots,u_a,v_1,\ldots,v_b,x}}\right] = [1-p^a(1-p)^b]^{n-a-b}.$$

Set $\epsilon = \min(p,1-p)^s$ so that

$$\Pr\left[\bigwedge_x \overline{E_{u_1,\ldots,u_a,v_1,\ldots,v_b,x}}\right] \leq (1-\epsilon)^{n-s}.$$

The key here is that ϵ is a fixed (dependent on p,s) positive number. Set

$$E = \bigvee E_{u_1,\ldots,u_a,v_1,\ldots,v_b},$$

the disjunction over all distinct $u_1,\ldots,u_a,v_1,\ldots,v_b \in G$ with $a+b \leq s$. There are less than $s^2n^s = O(n^s)$ such choices, as we can choose a,b and then the vertices. Thus

$$\Pr[E] \leq s^2n^s(1-\epsilon)^{n-s}.$$

But

$$\lim_{n\to\infty} s^2n^s(1-\epsilon)^{n-s} = 0$$

and so E holds almost never. Thus $\neg E$, which is precisely the statement that $G(n,p)$ has the full level s extension property, holds almost always. ■

But now we have proven Theorem 7.1. For any $p \in (0,1)$ and any fixed s as $m,n \to \infty$ with probability approaching 1, both $G(n,p)$ and $H(m,p)$ will have the full level s extension property and so Duplicator will win EHR[$G(n,p)$, $H(m,p),s$].

Why can't Duplicator use this strategy when $p = n^{-\alpha}$? We illustrate the difficulty with a simple example. Let $.5 < \alpha < 1$ and let Spoiler and Duplicator play a three-move game on G,H. Spoiler thinks of a point $z \in G$, but doesn't tell Duplicator about it. Instead, he picks $x_1,x_2 \in G$, both adjacent to z. Duplicator simply picks $y_1,y_2 \in H$, either adjacent or not adjacent dependent on whether $x_1 \sim x_2$. But now wily Spoiler picks $x_3 = z$. $H \sim H(m,m^{-\alpha})$ does not have the full level 2 extension property. In particular, most pairs y_1,y_2 do not have a common neighbor. Unless Duplicator is lucky, or shrewd, he then cannot find $y_3 \sim y_1,y_2$ and so he loses. This example does not say that Duplicator will lose with perfect play—indeed, we will show that he almost always wins with perfect play—it only indicates that the strategy used need be more complex. Now let us fix $\alpha \in (0,1)$, α irrational.

We define a *rooted graph* to be a pair (R,H) where H is a graph on vertex set, say, $V(H) = \{x_1,\ldots,x_r,y_1,\ldots,y_v\}$, and $R = \{x_1,\ldots,x_r\}$ is a specified subset of $V(H)$, called the roots. For example, (R,H) might consist of one vertex y_1 adjacent to the two roots x_1,x_2. Let $v = v(R,H)$ denote the number of vertices that are not roots and let $e = e(R,H)$ denote the number of edges, excluding those edges between two roots. We say (R,H) is *dense* if $v - e\alpha < 0$ and *sparse* if $v - e\alpha > 0$. The irrationality of α assures us that all (R,H) are in one of these categories. We call (R,H) *rigid* if for all S with $R \subseteq S \subset V(H)$, (S,H) is dense.

We think of rooted graphs as on abstract points. Suppose $V(H) = \{x_1,\ldots,x_r,y_1,\ldots,y_v\}$ and $R = \{x_1,\ldots,x_r\}$ as before. In a graph G, we say that vertices $Y_1,\ldots,Y_v$ form an (R,H) extension of $X_1,\ldots,X_r$ if whenever x_i is adjacent to y_j in H, X_i is adjacent to Y_j in G and also whenever y_i and y_j are adjacent in H, Y_i and Y_j are adjacent in G. Note that we allow G to have more edges than H and that the edges between the roots "don't count." Henceforth we will use lowercase letters for denoting vertices of H or of G. For any r,t, there is a finite list (up to isomorphism) of rigid rooted graphs (R,H) containing r roots and with $v(R,H) \le t$. In any graph G, we define the t-closure $cl_t(x_1,\ldots,x_r)$ to be the union of all $y_1,\ldots,y_v$ with (crucially) $v \le t$ which form an (R,H) extension where (R,H) is rigid. If there are no such sets, we define the default value $cl_t(x_1,\ldots,x_r) = \{x_1,\ldots,x_r\}$. We say two sets $x_1,\ldots,x_r$ and $x'_1,\ldots,x'_r$ have the same t-type if their t-closures are isomorphic. (To be precise, these are ordered r-tuples and the isomorphism must send x_i into x'_i.)

Example. Taking $\alpha \sim .51$ (but irrational, of course), $cl_1(x_1,x_2)$ consists of x_1,x_2 and all y adjacent to both of them. $cl_3(x_1,x_2)$ has those points and all y_1,y_2,y_3 which together with x_1 form a K_4 (note that this gives an (R,H) with $v = 3, e = 6$) and a finite number of other possibilities. ■

We can already describe the nature of Duplicator's strategy. At the end of the rth move, with $x_1,\ldots,x_r$ and $y_1,\ldots,y_r$ having been selected from the two graphs, Duplicator will assure that these sets have the same a_r*-type*. We shall call this the $(a_1,\ldots,a_t)$ *lookahead strategy*. Here a_r must depend only on t, the total number of moves in the game, and α. We shall set $a_t = 0$ so that at the end of the game, if Duplicator can stick to the $(a_1,\ldots,a_t)$ lookahead strategy, then he has won. If, however, Spoiler picks, say, x_r, so that there is no corresponding y_r with $x_1,\ldots,x_r$ and $y_1,\ldots,y_r$ having the same a_r-type, then the strategy fails and we say that Spoiler wins. The values a_r give the "lookahead" that Duplicator uses, but before defining them we need some preliminary results.

Lemma 7.6. *Let α, $r,t > 0$ be fixed. Then there exists $K = K(\alpha,r,t)$ so that in $G(n,n^{-\alpha})$ a.s.,*

$$|cl_t(x_1,\ldots,x_r)| \leq K$$

for all $x_1,\ldots,x_r \in G$.

Proof. Set $K = r + t(L-1)$. If $X = \{x_1,\ldots,x_r\}$ has t-closure with more than K points, then there will be L sets $Y^1,\ldots,Y^L$ disjoint from X, all $|Y^j| \leq t$, so that each $(X, X\cup Y^j)$ forms a rigid extension and with each Y^j having at least one point not in $Y^1 \cup \ldots Y^{j-1}$. Begin with X and add the Y^j in order. Adding Y^j will add, say, v_j vertices and e_j edges. Since $(X, X\cup Y^j)$ was *rigid*, $(X\cup Y^1 \cup\ldots\cup Y^{j-1}, X\cup Y^1\cup\ldots\cup Y^j)$ is dense and so $v_j - e_j\alpha < 0$. As $v_j \leq t$, there are only a finite number of possible values of $v_j - e_j\alpha$ and so there is an $\epsilon = \epsilon(\alpha,r,t)$ so that all $v_j - e_j\alpha \leq -\epsilon$. Pick L (and therefore K) so that $r - L\epsilon < 0$. The existence of a t-closure of size greater than K would imply the existence in $G(n,n^{-\alpha})$ of one of a finite number of graphs that would have some $r + v_1 + \cdots + v_L$ vertices and at least $e_1 + \cdots + e_L$ edges. But the probability of G containing such a graph is bounded by

$$\begin{aligned} n^{r+v_1+\cdots+v_L}p^{e_1+\cdots+e_L} &= n^{r+v_1+\cdots+v_L-\alpha(e_1+\cdots+e_L)} \\ &= n^{r+(v_1-\alpha e_1)+\cdots+(v_L-\alpha e_L)} \\ &\leq n^{r-L\epsilon} \\ &= o(1), \end{aligned}$$

so a.s. no such t-closures exist. ■

Remark. The value of K given by the above proof depends strongly on how close α may be approximated by rationals of denominator at most t. This is often the case. If, for example, $1/2 + 1/(s+1) < \alpha < 1/2 + 1/s$, then a.s. there will be two points $x_1, x_2 \in G(n,n^{-\alpha})$ having s common neighbors so that $|cl_1(x_1,x_2)| = s+2$.

Now we define the $a_1,\ldots,a_t$ of the lookahead strategy by reverse induction. We set $a_t = 0$. If at the end of the game Duplicator can assure that the 0-types of $x_1,\ldots,x_t$ and $y_1,\ldots,y_t$ are the same, then they have the same induced subgraphs and he has won. Suppose, inductively, that $b = a_{r+1}$ has been defined. Let, applying the lemma, K be a.s. an upper bound on all $cl_b(z_1,\ldots,z_{r+1})$. We then define $a = a_r$ by $a = K + b$.

Now we need show that a.s. this strategy works. Let $G_1 \sim G(n,n^{-\alpha})$, $G_2 \sim G(m,m^{-\alpha})$ and suppose Duplicator tries to play the $(a_1,\ldots,a_t)$ lookahead strategy on EHR(G_1,G_2,t).

Set $a = a_1$ and consider the first move. Spoiler will select, say, $y = y_1 \in G_2$. Duplicator then must play $x = x_1 \in G_1$ with $cl_a(x) \cong cl_a(y)$. Can he always do so—that is, do a.s. G_1 and G_2 have the same values of $cl_a(x)$? The size of $cl_a(x)$ is a.s. bounded, so it suffices to show for any potential H that either there almost surely is an x with $cl_a(x) \cong H$ or there almost surely is no x with $cl_a(x) \cong H$.

Let H have v vertices and e edges. Suppose H has a subgraph H' (possibly H itself) with v' vertices, e' edges, and $v' - \alpha e' < 0$. The expected number of copies of H' in G_1 is

$$\Theta(n^{v'}p^{e'}) = \Theta(n^{v'-\alpha e'}) = o(1),$$

so a.s. G_1 contains no copy of H', hence no copy of H, hence no x with $cl_a(x) \cong H$. If this does not occur then (since, critically, α is irrational) all $v' - \alpha e' > 0$, so the expected number of copies of all such H' approaches infinity. Applying Theorem 4.4 of Chapter 4, a.s. G_1 has $\Theta(n^{v-\alpha e})$ copies of H. For x in appropriate position in such a copy of H, we cannot deduce $cl_a(x) \cong H$ but only that $cl_a(x)$ contains H as a subgraph. (Essentially, x may have additional extension properties.) For each such x, as $cl_a(x)$ is bounded, $cl_a(x)$ contains only a bounded number of copies of H. Hence there are $\Theta(n^{v-\alpha e})$ *different* $x \in G_1$, so that $cl_a(x)$ contains H as a subgraph.

Let H' be a possible value for $cl_a(x)$ that contains H as a proper subgraph. Let H' have v' vertices and e' edges. As (x,H') is rigid, (H,H') is dense and so

$$(v' - v) - \alpha(e' - e) < 0.$$

There are $\Theta(n^{v'-\alpha e'})$ different x with $cl_a(x)$ containing H', but since $v' - \alpha e' < v - \alpha e$, this is $o(n^{v-\alpha e})$. Deleting all such x for all such H', there a.s. remain $\Theta(n^{v-\alpha e})$, and hence at least one, x with $cl_a(x) \cong H$.

Now, in general, consider the $(r+1)$ move. We set $b = a_{r+1}$, $a = a_r$ for notational convenience and recall $a = K + b$, where K is an upper bound on $cl_b(z_1,\ldots,z_{r+1})$. Points $x_1,\ldots,x_r \in G_1$, $y_1,\ldots,y_r \in G_2$ have been selected with

$$cl_a(x_1,\ldots,x_r) \cong cl_a(y_1,\ldots,y_r).$$

Spoiler picks, say, $x_{r+1} \in G_1$. We distinguish two cases. We say Spoiler has moved inside if

$$x_{r+1} \in cl_K(x_1,\ldots,x_r).$$

Otherwise we say Spoiler has moved outside.

Suppose Spoiler moves inside. Then

$$cl_b(x_1,\ldots,x_r,x_{r+1}) \subseteq cl_{K+b}(x_1,\ldots,x_r) = cl_a(x_1,\ldots,x_r).$$

The isomorphism from $cl_a(x_1,\ldots,x_r)$ to $cl_a(y_1,\ldots,y_r)$ sends x_{r+1} to some y_{r+1} that Duplicator selects.

Suppose Spoiler moves outside. Set $H = cl_b(x_1,\ldots,x_r,x_{r+1})$. Let H_0 be the union of all rigid extensions of any size of $x_1,\ldots,x_r$ in H. If $x_{r+1} \in H_0$, then, as $|H| \leq K$, $x_{r+1} \in cl_K(x_1,\ldots,x_r)$ and Spoiler actually moved inside. Hence $x_{r+1} \notin H_0$. Since $|H| \leq K \leq a$, H_0 lies inside $cl_a(x_1,\ldots,x_r)$. The isomorphism between $cl_a(x_1,\ldots,x_r)$ and $cl_a(y_1,\ldots,y_r)$ maps H_0 into a copy of itself in the graph G_2.

For any copy of H_0 in G_2, let $N(H_0)$ denote the number of extensions of H_0 to H. From the methods of Chapter 8, Section 5, one can show that a.s all $N(H_0) = \Theta(n^{v-\alpha e})$, with $v = v(H_0,H)$, $e = e(H_0,H)$, and $v - \alpha e > 0$. (The methods used in proving Theorem 5.3 in the aforementioned section apply in this more general case.) For a given H_0, each y_{r+1} is in only a bounded number of copies of H since all copies of H lie in $cl_b(y_1,\ldots,y_r,y_{r+1})$. Hence there are $\Theta(n^{v-\alpha e})$ vertices y_{r+1} so that $cl_b(y_1,\ldots,y_r,y_{r+1})$ contains H. Arguing as with the first move there a.s. are $\Theta(n^{v-\alpha e})$, hence at least one, y_{r+1} with $cl_b(y_1,\ldots,y_r,y_{r+1}) \cong H$. Duplicator selects such a y_{r+1}.

The Probabilistic Lens: Counting Subgraphs

A graph $G=(V,E)$ on n vertices has 2^n induced subgraphs, but some will surely be isomorphic. How many different subgraphs can G have? Here we show that there are graphs G with $2^n(1-o(1))$ different subgraphs. The argument we give is fairly coarse. It is typical of those situations where a probabilistic approach gives fairly quick answers to questions otherwise difficult to approach.

Let G be a random graph on n vertices with edge probability $1/2$. Let $S \subseteq V$, $|S|=t$, be fixed. For any one-to-one $\rho : S \to V$, $\rho \neq id$, let A_ρ be the event that ρ gives a graph isomorphism—that is, for $x,y \in S$, $\{x,y\} \in E \Leftrightarrow \{\rho x, \rho y\} \in E$. Set $M_\rho = \{x \in S : \rho x \neq x\}$. We split the set of ρ by $g = g(\rho) = |M_\rho|$.

Consider the $g(t-g)+\binom{g}{2}$ pairs x,y with $x,y \in S$ and at least one of x,y in M. For all but at most $g/2$ of these pairs, $\{x,y\} \neq \{\rho x, \rho y\}$. (The exceptions are when $\rho x = y, \rho y = x$.) Let E_ρ be the set of pairs $\{x,y\}$ with $\{x,y\} \neq \{\rho x, \rho y\}$. Define a graph H_ρ with vertices E_ρ and vertex $\{x,y\}$ adjacent to $\{\rho x, \rho y\}$. In H_ρ, each vertex has degree at most 2 ($\{x,y\}$ may also be adjacent to $\{\rho^{-1}x, \rho^{-1}y\}$), and so it decomposes into isolated vertices, paths, and circuits. On each such component, there is an independent set of size at least one-third the number of elements, the extreme case being a triangle. Thus there is a set $I_\rho \subseteq E_\rho$ with

$$|I_\rho| \geq |E_\rho| \geq \frac{g(t-g)+\binom{g}{2}-g/2}{3},$$

so that the pairs $\{x,y\},\{\rho x, \rho y\}$ with $\{x,y\} \in I_\rho$ are all distinct.

For each $\{x,y\} \in I_\rho$, the event $\{x,y\} \in E \Leftrightarrow \{\rho x, \rho y\} \in E$ has probability $1/2$. Moreover, these events are mutually independent over $\{x,y\} \in I_\rho$, since they involve distinct pairs. Thus we bound

$$\Pr[A_\rho] \leq 2^{-|I_\rho|} \leq 2^{-(g(t-g)+\binom{g}{2}-g/2)/3}.$$

For a given g, the function ρ is determined by $\{x : \rho x \neq x\}$ and the values ρx for those x so that there are less than n^{2g} such ρ. We bound

$$\sum_{\rho \neq id} \Pr[A_\rho] = \sum_{g=1}^{t} \sum_{g(\rho)=g} \Pr[A_\rho] \leq \sum_{g=1}^{t} n^{2g} 2^{-(g(t-g)+\binom{g}{2}-g/2)/3}.$$

We make the rough bound

$$g(t-g)+\binom{g}{2}-\frac{g}{2} = g\left(t-\frac{g}{2}-1\right) \geq g\left(\frac{t}{2}-1\right),$$

since $g \leq t$. Then

$$\sum_{\rho \neq id} \Pr[A_\rho] \leq \sum_{g=1}^{t} \left\{n^2 2^{[(-t/2)+1]/3}\right\}^g.$$

For, again being rough, $t > 50 \ln n$, $n^2 2^{-t/6} < n^{-3}$ and $\sum_{\rho \neq id} \Pr[A_\rho] = o(1)$. That is, almost surely there is no isomorphic copy of $G|_S$.

For all $S \subseteq V$ with $|S| > 50 \ln n$, let I_S be the indicator random variable for there being no other subgraph isomorphic to $G|_S$. Set $X = \sum I_S$. Then $E[I_S] = 1 - o(1)$, so, by linearity of expectation—there being $2^n(1-o(1))$ such S—

$$E[X] = 2^n(1-o(1)).$$

Hence there is a specific G with $X > 2^n(1-o(1))$.

11

Circuit Complexity

1. PRELIMINARIES

A *Boolean function* $f = f(x_1, \ldots, x_n)$ on the n variables $x_1, x_2, \ldots, x_n$ is simply a function $f : \{0,1\}^n \to \{0,1\}$. In particular, $0, 1, x_1 \wedge \cdots \wedge x_n$, $x_1 \vee \cdots \vee x_n$, $x_1 \oplus \cdots \oplus x_n$ denote, as usual, the two constant functions, the *And* function (whose value is 1 iff $x_i = 1$ for all i), the *Or* function (whose value is 0 iff $x_i = 0$ for all i), and the *parity* function (whose value is 0 iff an even number of variables x_i is 1), respectively. For a function f, we let $\overline{f} = f \oplus 1$ denote its complement, *not* f. The functions x_i and $\overline{x}_i$ are called *atoms*. In this section, we consider the problem of computing various Boolean functions efficiently. A *circuit* is a directed, acyclic graph, with a special vertex with no outgoing edges called the output vertex. Every vertex is labeled by a Boolean function of its immediate parents, and the vertices with no parents (i.e., those with no ingoing edges) are labeled either by one of the variables x_i or by a constant 0 or 1. For every assignment of binary values to each variable x_i, one can compute, recursively, the corresponding value of each vertex of the circuit by applying the corresponding function labeling it to the already computed values of its parents. We say that the circuit *computes* the function $f = f(x_1, \ldots, x_n)$ if for each $x_i \in \{0,1\}$, the corresponding value of the output vertex of the circuit equals $f(x_1, \ldots, x_n)$. For example, Figure 1 presents a circuit computing $f(x_1, x_2, x_3) = (x_1 \oplus (x_2 \wedge x_3)) \wedge x_1$.

If every fanout in a circuit is at most 1 (i.e., the corresponding graph is a tree), the circuit is called a *formula*. If every fanin in a circuit is at most 2, the circuit is called a *binary circuit*. Therefore the circuit in Figure 1 is binary, but it is not a formula. The *size* of a circuit is the number of vertices in it, and its *depth* is the maximum length of a directed path in it. The *binary circuit complexity* of a Boolean function is the size of the smallest binary circuit computing it. An easy counting argument shows that for large n, the binary circuit complexity of almost all the functions of n variables is at least $(1 + o(1))2^n/n$. This is because the number of binary circuits of size s can be easily shown to be less than $(c_1 s)^s$, whereas the total number of Boolean functions on n variables is 2^{2^n}. On the other hand, there is no known nonlinear, not to mention exponential (in n), lower bound for the binary circuit complexity of any "explicit" function. By "explicit" here, we mean an NP function,

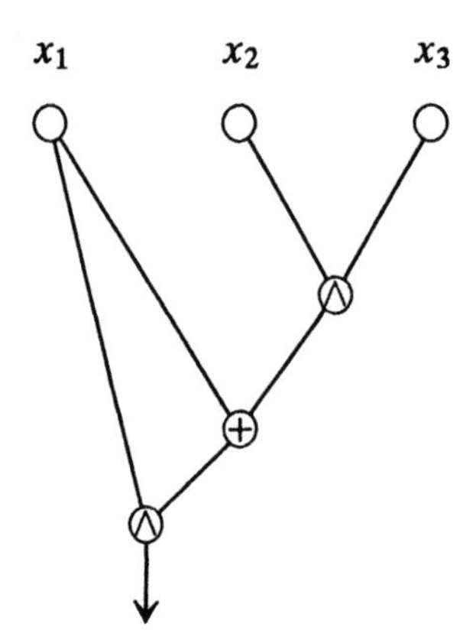

Figure 1

that is, one of a family $\{f_{n_i}\}_{i\geq 1}$ of Boolean functions, where f_{n_i} has n_i variables, $n_i \to \infty$, and there is a nondeterministic Turing machine which, given n_i and $x_1,\ldots,x_{n_i}$, can decide in (nondeterministic) polynomial time (in n_i) if $f_{n_i}(x_1,\ldots,x_{n_i}) = 1$. (An example for such a family is the $n/2$-clique function; here $n_i = \binom{i}{2}$, the n_i variables $x_1,\ldots,x_{n_i}$ represent the edges of a graph on i vertices and $f_{n_i}(x_1,\ldots,x_{n_i}) = 1$ iff the corresponding graph contains a clique on at least $i/2$ vertices.) Any nonpolynomial lower bound for the binary circuit complexity of an explicit function would imply (among other things) that $P \neq NP$ and thus solve the arguably most important open problem in theoretical computer science. Unfortunately, at the moment, the best known lower bound for the binary circuit complexity of an explicit function of n variables is only $3n$ (see Blum [1984]; Paul [1977]). However, several nontrivial lower bounds are known when we impose certain restrictions on the structure of the circuits. Most of the known proofs of these bounds rely heavily on probabilistic methods. In this chapter, we describe some of these results. We note that there are many additional beautiful known results about circuit complexity—see, e.g., Wegener (1987) and Krachmer and Wigderson (1990)—but those included here are not only among the crucial ones, but also represent the elegant methods used in this field. Since most results in this chapter are asymptotic, we assume, throughout the chapter, whenever it is needed, that the number of variables we have is sufficiently large.

2. RANDOM RESTRICTIONS AND BOUNDED-DEPTH CIRCUITS

Let us call a Boolean function G a t-And–Or if it can be written as an And of an arbitrary number of functions, each being an Or of at most t atoms, i.e., $G = G_1 \wedge \cdots \wedge G_w$, where $G_i = y_{i1} \vee \cdots \vee y_{ia_i}$, $a_i \leq t$, and each y_j is an atom. Similarly, we call a Boolean function an s-Or–And if it can be written as an Or of And gates each containing at most s atoms. A *minterm* of a function is a minimal assignment of values to some of the variables that forces the function to be 1. Its *size* is the number of variables whose values are set. Notice that a function is an s-Or–And if and only if each of its minterms is of size at most s. A *restriction* is a map ρ of the set of indices $\{1,\dots,n\}$ to the set $\{0,1,*\}$. The restriction of the function $G = G(x_1,\dots,x_n)$ by ρ, denoted by $G \mid \rho$, is the Boolean function obtained from G by setting the value of each x_i for $i \in \rho^{-1}\{0,1\}$ to $\rho(i)$, and leaving each x_j for $j \in \rho^{-1}(*)$ as a variable. Thus, for example, if $G(x_1,x_2,x_3) = (x_1 \wedge x_2) \vee x_3$ and $\rho(1) = 0$ $\rho(2) = \rho(3) = *$, then $G\,|_\rho = x_3$. For $0 \le p \le 1$, a *random p-restriction* is a random restriction ρ defined by choosing, for each $1 \le i \le n$ independently, the value of $\rho(i)$ according to the following distribution:

$$\Pr[\rho(i) = *] = p, \qquad \Pr[\rho(i) = 0] = \Pr[\rho(i) = 1] = \frac{(1-p)}{2}. \tag{1}$$

Improving results of Furst, Saxe, and Sipser (1984), Ajtai (1983), and Yao (1985), Hastad (1988) proved the following result, which is very useful in establishing lower bounds for bounded-depth circuits.

Lemma 2.1 (The Switching Lemma). *Let $G = G(x_1,\dots,x_n)$ be a t-And–Or, i.e., $G = G_1 \wedge G_2 \wedge \cdots \wedge G_w$, where each G_i is an Or of at most t atoms. Let ρ be the random restriction defined by* (1). *Then*

$$\Pr[G \mid \rho \text{ is not an } (s-1)\text{-Or–And}]$$
$$= \Pr[G \mid \rho \text{ has a minterm of size } \ge s] \le (5pt)^s.$$

Proof. Let E_s be the event that $G\,|_\rho$ has a minterm of size at least s. To bound $\Pr(E_s)$, we prove a stronger result; for any Boolean function F,

$$\Pr[E_s \mid F\,|_\rho \equiv 1] \le (5pt)^s. \tag{2}$$

Here we agree that if the condition is unsatisfied, then the conditional probability is 0. Lemma 2.1 is obtained from (2) by taking $F \equiv 1$. We prove (2) by induction on w. For $w = 0$, $G \equiv 1$ and there is nothing to prove. Assuming (2) holds whenever the number of G_i is less than w, we prove it for w. Put $G = G_1 \wedge G^*$, where $G^* = G_2 \wedge \cdots \wedge G_w$, and let E_s^* be the event that $G^*\,|_\rho$ has a minterm of size at least s. By interchanging, if necessary, some of the variables with their complements, we may assume, for convenience, that $G_1 = \bigvee_{i \in T} x_i$, where $|T| \le t$. Either $G_1\,|_\rho \equiv 1$ or $G_1\,|_\rho \not\equiv 1$. In the former case, E_s holds if and only if E_s^* holds and hence, by induction

$$\Pr[E_s \mid F\,|_\rho \equiv 1,\ G_1\,|_\rho \equiv 1] = \Pr[E_s^* | (F \wedge G_1)|_\rho \equiv 1] \le (5pt)^s. \tag{3}$$

The case $G_1\,|_\rho \not\equiv 1$ requires more work. In this case, any minterm of $G\,|_\rho$ must assign a value 1 to at least one x_i, for $i \in T$. For a nonempty $Y \subseteq T$ and for a function $\sigma : Y \to \{0,1\}$ that is not identically 0, let $E_s(Y,\sigma)$ be the event that $G\,|_\rho$ has a minterm of size at least s which assigns the value $\sigma(i)$ to x_i

for each $i \in Y$ and does not assign any additional values to variables x_j with $j \in T$. By the preceding remark,

$$\Pr[E_s \mid F|_\rho \equiv 1,\ G_1|_\rho \not\equiv 1] \le \sum_{Y,\sigma} \Pr[E_s(Y,\sigma) \mid F|_\rho \equiv 1,\ G_1|_\rho \not\equiv 1]. \tag{4}$$

Observe that the condition $G_1|_\rho \not\equiv 1$ means precisely that $\rho(i) \in \{0,*\}$ for all $i \in T$ and hence, for each $i \in T$,

$$\Pr[\rho(i) = * \mid G_1|_\rho \not\equiv 1] = \frac{p}{p + (1-p)/2} = \frac{2p}{1+p}.$$

Thus, if $|Y| = y$,

$$\Pr[\rho(Y) = * \mid G_1|_\rho \not\equiv 1] \le \left(\frac{2p}{1+p}\right)^y.$$

The further condition $F|_\rho \equiv 1$ can only decrease this probability. This can be shown using the FKG inequality (see Chapter 6). It can also be shown directly as follows. For any fixed $\rho' : N - Y \to \{0,1,*\}$, where $N = \{1,\ldots,n\}$, we claim that

$$\Pr[\rho(Y) = * \mid F|_\rho \equiv 1,\ G_1|_\rho \not\equiv 1,\ \rho|_{N-Y} = \rho'] \le \left(\frac{2p}{1+p}\right)^y.$$

Indeed, the given ρ' has a unique extension ρ with $\rho(Y) = *$. If that ρ does not satisfy the above conditions, then the conditional probability is 0. If it does, then so do all extensions ρ with $\rho(i) \in \{0,*\}$ for $i \in Y$, and so the inequality holds in this case too. As this holds for all fixed ρ', we conclude that indeed

$$\Pr[\rho(Y) = * \mid F|_\rho \equiv 1,\ G_1|_\rho \not\equiv 1] \le \left(\frac{2p}{1+p}\right)^y \le (2p)^y. \tag{5}$$

Let $\rho' : T \to \{0,*\}$ satisfy $\rho(Y) = *$ and consider all possible restrictions ρ satisfying $\rho|_T = \rho'$. Under this condition, ρ may be considered as a random restriction on $N - T$. The event $F|_\rho \equiv 1$ reduces to the event $\tilde{F}|_{\rho|_{N-T}} \equiv 1$, where $\tilde{F}$ is the And of all functions obtained from F by substituting the values of x_i according to ρ' for those $i \in T$ with $\rho'(i) = 0$, and by taking all possibilities for all the other variables x_j for $j \in T$. If the event $E_s(Y,\sigma)$ occurs, then $G^*|_{\rho\sigma}$ has a minterm of size at least $s - y$ that does not contain any variable x_i with $i \in T - Y$. But this happens if and only if $\tilde{G}|_{\rho|_{N-T}}$ has a minterm of size at least $s - y$, where $\tilde{G}$ is the function obtained from G^* by substituting the values of x_j for $j \in Y$ according to σ, the values of x_i for $i \in T - Y$ and $\rho'(i) = 0$ according to ρ', and by removing all the variables x_k with $k \in T - Y$ and $\rho'(k) = *$. Denoting this event by $\tilde{E}_{s-y}$ we can apply induction and obtain

$$\Pr[E_s(Y,\sigma) | F|_\rho \equiv 1,\ G_1|_\rho \not\equiv 1,\ \rho|_T = \rho'] \le \Pr[\tilde{E}_{s-y} \mid \tilde{F}|_\rho \equiv 1] \le (5pt)^{s-y}.$$

Since any ρ with $F|_\rho \equiv 1$, $G_1|_\rho \equiv 1$, $\rho(Y) = *$ must have $\rho|_T = \rho'$ for some ρ' of this form, and since the event $E_s(Y,\sigma)$ may occur only if $\rho(Y) = *$, we conclude that

$$\Pr[E_s(Y,\sigma) \mid F\mid_\rho \equiv 1,\ G_1\mid_\rho \not\equiv 1,\ \rho(Y) = *] \le (5pt)^{s-y},$$

and, by (5),

$$\begin{aligned}\Pr[E_s(Y,\sigma) \mid F\mid_\rho \equiv 1,\ G_1\mid_\rho \not\equiv 1] &= \Pr[\rho(Y) = * \mid F\mid_\rho \equiv 1,\ G_1\mid_\rho \not\equiv 1]\\ &\cdot \Pr[E_s(Y,\sigma) \mid F\mid_\rho \equiv 1,\ G_1\mid_\rho \not\equiv 1,\ \rho(Y) = *]\\ &\le (2p)^y(5pt)^{s-y}.\end{aligned}$$

Substituting in (4) and using the fact that $|T| \le t$ and that

$$\sum_{y=1}^{t} \frac{(2^y-1)2^y}{(5^y y!)} \le \frac{2}{5} + \sum_{y=2}^{\infty} \frac{(4/5)^y}{y!} = \frac{2}{5} + e^{4/5} - 1 - \frac{4}{5} < 1,$$

we obtain,

$$\begin{aligned}\Pr[E_s \mid F\mid_\rho \equiv 1,\ G_1\mid_\rho \not\equiv 1] &\le \sum_{y=1}^{|T|} \binom{|T|}{y} (2^y - 1)(2p)^y(5pt)^{s-y}\\ &\le (5pt)^s \sum_{y=1}^{t} \frac{t^y}{y!}(2^y-1)\left(\frac{2}{5t}\right)^y\\ &= (5pt)^s \sum_{y=1}^{t} (2^y - 1)\cdot \frac{2^y}{5^y \cdot y!} \le (5pt)^s.\end{aligned}$$

This, together with (3), gives

$$\Pr[E_s \mid F\mid_\rho \equiv 1] \le (5pt)^s,$$

completing the induction and the proof. ∎

By taking the complement of the function G in Lemma 2.1 and applying De Morgan's rules, one clearly obtains its dual form: if G is a t-Or–And and ρ is the random restriction given by (1), then $\Pr[G\mid_\rho$ is not an $(s-1)$-And–Or$] \le (5pt)^s$.

We now describe an application of the switching lemma that supplies a lower bound to the size of circuits of small depth that compute the parity function $x_1 \oplus \cdots \oplus x_n$. We consider circuits in which the vertices are arranged in levels, those in the first level are atoms (i.e., variables or their complements), and each other gate is either an Or or an And of an arbitrary number of vertices from the previous level. We assume that the gates in each level are either all And gates or all Or gates, and that the levels alternate between And levels and Or levels. A circuit of this form is called a $C(s,s',d,t)$-circuit if it contains at most s gates, at most s' of which are above the second level, its depth is at most d, and the fanin of each gate in its second level is at most t.

Thus, for example, the circuit that computes the parity function by computing an Or of the 2^{n-1} terms $x_1^{\epsilon_1} \wedge \cdots \wedge x_n^{\epsilon_n}$, where $(\epsilon_1, \ldots, \epsilon_n)$ ranges over all even binary vectors and $x_i^{\epsilon_i} = x_i \oplus \epsilon_i$, is a $C(2^{n-1}+1, 1, 2, n)$-circuit.

Theorem 2.2. *Let $f = f(x_1, \ldots, x_n)$ be a function and let C be a $C(\infty, s, d, t)$-circuit computing f, where $s \cdot (1/2)^t \leq 0.5$. Then either f or its complement $\overline{f}$ has a minterm of size at most $n - n/[2 \cdot (10t)^{d-2}] + t$.*

Proof. Let us apply to C, repeatedly, $d-2$ times a random $1/(10t)$-restriction. Each of these random restrictions, when applied to any bottom subcircuit of depth 2, transforms it by Lemma 2.1 with probability at least $1-(1/2)^t$ from a t-Or–And to a t-And–Or (or conversely). If all these transformations succeed, we can merge the new And gates with these from the level above them and obtain a circuit with a smaller depth. As the total size of the circuit is at most s, and $s(1/2)^t \leq .5$, we conclude that with probability at least half all transformations succeed and C is transformed into a $C(\infty, 1, 2, t)$-circuit. Each variable x_i, independently, is still a variable (i.e., has not been assigned a value) with probability $1/(10t)^{d-2}$. Thus the number of remaining variables is a binomial random variable with expectation $n/(10t)^{d-2}$ and a little smaller variance. By the standard estimates for binomial distributions (see Appendix A), the probability that at least $n/[2 \cdot (10t)^{d-2}]$ variables are still variables is more than a half. Therefore, with positive probability, at most $n - n/[2 \cdot (10t)^{d-2}]$ of the variables have been fixed and the resulting restriction of f has a $C(\infty, 1, 2, t)$-circuit, i.e., its value can be fixed by assigning values to at most t additional variables. This completes the proof. ∎

Corollary 2.3. *For any $d \geq 2$, there is no*

$$C\left(\infty, \tfrac{1}{2} \cdot 2^{1/10[n^{1/(d-1)}]}, d, \tfrac{1}{10} n^{1/(d-1)}\right)\text{-circuit}$$

that computes the parity function $f(x_1, \ldots, x_n) = x_1 \oplus \cdots \oplus x_n$.

Proof. Assuming there is such a circuit, we obtain, by Theorem 2.2, that the value of f can be fixed by assigning values to at most $n - \frac{1}{2} n^{1/(d-1)} + \frac{1}{10} n^{1/(d-1)} < n$ variables. This is false, and hence there is no such circuit. ∎

The estimate in Corollary 2.3 is, in fact, nearly best possible. Since every $C(s, s', d, t)$-circuit can be transformed into a $C(ts, s, d+1, 2)$-circuit (by replacing each atom by an Or or And of two copies of itself), Corollary 2.3 easily implies that the depth d of any $C(s, s', d, t)$-circuit of polynomial size is at least $\Omega(\log n / \log\log n)$. This lower bound is also optimal.

3. MORE ON BOUNDED-DEPTH CIRCUITS

In the previous section, we saw that the parity function is hard to compute in small depth using And, Or, and Not gates. It turns out that even if we allow the use of parity gates (in addition to the And, Or, and Not gates), there are still some relatively simple functions that are hard to compute. Such a result was first proved by Razborov (1987). His method was modified and strengthened by Smolensky (1987). For an integer $k \geq 2$, let $\mathrm{Mod}_k(x_1, x_2, \ldots, x_n)$ be the Boolean function whose value is 1 iff $\Sigma x_i \not\equiv 0 (\mathrm{mod}\, k)$. Smolensky showed that for every two powers p and q of distinct primes, the function Mod_p cannot be computed in a bounded-depth polynomial-size circuit that uses And, Or, Not, and Mod_q gates. Here we present the special case of this result, in which $q = 3$ and $p = 2$.

Let C be an arbitrary circuit of depth d and size s consisting of And, Or, Not, and Mod_3 gates. A crucial fact, due to Razborov, is the assertion that the output of C can be approximated quite well (depending on d and s) by a polynomial of relatively small degree over $GF(3)$. This is proved by applying the probabilistic method as follows. Let us replace each gate of the circuit C by an approximate polynomial operation, according to the following three rules, which guarantee that in each vertex in the new circuit we compute a polynomial over $GF(3)$, whose values are all 0 or 1 (whenever the input is a 0–1 input).

(i) Each Not-gate $\overline{y}$ is replaced by the polynomial gate $(1 - y)$.

(ii) Each Mod_3 gate $\mathrm{Mod}_3(y_1, \ldots, y_m)$ is replaced by the polynomial gate $(y_1 + y_2 + \cdots + y_m)^2$.

The rule for replacement of Or and And gates is a little more complicated. Observe that in the two previous cases, (i) and (ii), there was no approximation; the new gates compute precisely what the old ones did, for all possible Boolean values of the variables. This can, in principle, be done here too. An And gate $y_1 \wedge \cdots \wedge y_m$ should simply be replaced by the product $y_1 \ldots y_m$. An Or gate $y_i \vee \ldots \vee y_m$ can then be computed by De Morgan's rules. Since $y_1 \vee \cdots \vee y_m = \overline{(\overline{y}_1 \wedge \cdots \wedge \overline{y}_m)}$ and $\overline{y}$ is realized by $(1 - y)$, this would give

$$1 - (1 - y_1)(1 - y_2) \ldots (1 - y_m). \tag{6}$$

The trouble is that this procedure would increase the degree of our polynomials too much. Thus, we need to be a little more tricky. Let ℓ be an integer, to be chosen later. Given an Or gate $y_1 \vee \cdots \vee y_m$, we choose ℓ random subsets $I_1, \ldots, I_\ell$ of $\{1, \ldots, m\}$, where for each $1 \leq i \leq \ell$ and for each $1 \leq j \leq m$ independently, $\Pr(j \in I_i) = 1/2$. Observe that for each fixed i, $1 \leq i \leq \ell$, the sum $(\sum_{j \in I_i} y_j)^2$ over $GF(3)$ is certainly 0 if $y_1 \vee \cdots \vee y_m = 0$, and is 1 with probability at least 1/2 if $y_1 \vee \cdots \vee y_m = 1$. Hence, if we compute the Or function of the ℓ expressions $(\sum_{j \in I_i} y_j)^2$ $(1 \leq i \leq \ell)$, this function is 0 if $y_1 \vee \cdots \vee y_m = 0$ and is 1 with probability at least $1 - (1/2)^\ell$ if $y_1 \vee \cdots \vee y_m = 1$. We thus compute the Or and write it as a polynomial, in the way explained in equation (6).

This gives

$$1-\prod_{i=1}^{\ell}\left(1-\left(\sum_{j\in I_i} y_j\right)^2\right). \tag{7}$$

(iii) We replace each Or gate by an approximation polynomial gate of the form described in (7). Once we have an approximation to an Or gate, we can obtain the corresponding one for an And gate by applying De Morgan rules. Since $y_1 \wedge \cdots \wedge y_m = \overline{(\overline{y}_1 \vee \cdots \vee \overline{y}_m)}$, we replace each And gate of the form $y_1 \wedge \cdots \wedge y_m$ by

$$\prod_{i=1}^{\ell}\left(1-\left[\sum_{j\in I_i}(1-y_j)\right]^2\right). \tag{8}$$

Observe that the polynomials in (6) and (7) are both of degree at most 2ℓ.

Given the original circuit C of depth d and size s, we can now replace all its gates by our approximative polynomial gates and get a new circuit CP, which depends on all the random choices made in each replacement of each of the And/Or gates. The new circuit CP computes a polynomial $P(x_1,\ldots,x_n)$ of degree at most $(2\ell)^d$. Moreover, for the fixed Boolean values of $x_1, x_2, \ldots, x_n$, the probability that all the new gates compute exactly what the corresponding gates in C computed is at least $1-s/2^\ell$. Therefore the expected number of inputs on which $P(x_1,\ldots,x_n)$ is equal to the output of C is at least $2^n(1-s/2^\ell)$. We have thus proved the following.

Lemma 3.1. *For any circuit C of depth d and size s on n Boolean variables that uses Not, Or, And, and* Mod$_3$ *gates and for any integer ℓ, there is a polynomial $P = P(x_1,\ldots,x_n)$ of degree at most $(2\ell)^d$ over $GF(3)$ whose value is equal to the output of C on at least $2^n(1-s/2^\ell)$ inputs.*

In order to apply this lemma for obtaining lower bounds for the size of any circuit of the above type that computes the parity function, we need the following additional combinatorial result.

Lemma 3.2. *There is no polynomial $P(x_1,\ldots,x_n)$ over $GF(3)$ of degree at most $\sqrt{n}$ that is equal to the parity of $x_1,\ldots,x_n$ for a set S of at least $.9\cdot 2^n$ distinct binary vectors $(x_1,\ldots,x_n)$.*

Proof. Suppose this is false, and suppose $S \subset \{0,1\}^n$, $|S| \geq .9\cdot 2^n$, and $P(x_1,\ldots,x_n) = x_1 \oplus \cdots \oplus x_n$ for all $(x_1,\ldots,x_n) \in S$. Define a polynomial $Q = Q(y_1,\ldots,y_n)$ by $Q = Q(y_1,\ldots,y_n) = P(y_1+2,\ldots,y_n+2)-2$, and $T = \{(y_1,\ldots,y_n) \in \{1,-1\}^n : (y_1+2,\ldots,y_n+2) \in S\}$, where all additions are in $GF(3)$. Clearly, Q has degree at most $\sqrt{n}$ and $Q(y_1,\ldots,y_n) = \prod_{i=1}^n y_i$ for all $(y_1,\ldots,y_n)$

$\in T$. Let now $G = G(y_1,\ldots,y_n) : T \to GF(3)$ be an arbitrary function. Extend it in an arbitrary way to a function from $(GF(3))^n \to GF(3)$, and write this function as a polynomial in n variables. (Trivially, any function from $(GF(3))^n \to GF(3)$ is a polynomial. This follows from the fact that it is a linear combination of functions of the form $\prod_{i=1}^{n}(y_i - \epsilon_i)(y_i - \epsilon_i - 1)$, where $\epsilon_i \in GF(3)$.) Replace each occurrence of y_i^2 in this polynomial by 1 to obtain a multilinear polynomial $\tilde{G}$ that agrees with G on T. Now replace each monomial $\prod_{i\in U} y_i$, where $|U| > n/2 + \sqrt{n}/2$ by $\prod_{i\notin U} y_i \cdot Q(y_1,\ldots,y_n)$, and replace this new polynomial by a multilinear one, $\tilde{\tilde{G}}$ again by replacing each y_i^2 by 1. Since for $y_i \in \{\pm 1\}$

$$\prod_{i\notin U} y_i \cdot \prod_{i=1}^{n} y_i = \prod_{i\in U} y_i,$$

$\tilde{\tilde{G}}$ is equal to G on T and its degree is at most $n/2 + \sqrt{n}/2$. However, the number of possible $\tilde{\tilde{G}}$ is

$$3^{\sum_{i=0}^{n/2+\sqrt{n}/2}\binom{n}{i}} < 3^{.8\cdot 2^n},$$

whereas the number of possible G is $3^{|T|} \geq 3^{.9\cdot 2^n}$. This is impossible, and hence the assertion of the lemma holds. ■

Corollary 3.3. *There is no circuit of depth d and size*

$$s \leq \tfrac{1}{10} 2^{n^{1/2d}/2}$$

computing the parity of $x_1, x_2, \ldots, x_n$ *using Not, And, Or, and* Mod$_3$ *gates.*

Proof. Suppose this is false and let C be such a circuit. Put $\ell = 1/2 \cdot n^{1/2d}$. By Lemma 3.1 there is a polynomial $P = P(x_1,\ldots,x_n)$ over $GF(3)$, whose degree is at most $(2\ell)^d = \sqrt{n}$, which is equal to the parity of $x_1,\ldots,x_n$ on at least

$$2^n\left(1 - \frac{s}{2^{n^{1/2d}/2}}\right) \geq .9 \cdot 2^n$$

inputs. This contradicts Lemma 3.2, and hence completes the proof. ■

4. MONOTONE CIRCUITS

A Boolean function $f = f(x_1,\ldots,x_n)$ is *monotone* if $f(x_1,\ldots,x_n) = 1$ and $x_i \leq y_i$ implies $f(y_1,\ldots,y_n) = 1$. A *binary monotone circuit* is a binary circuit that contains only binary And and Or gates. It is easy to see that a function is monotone if and only if there is a binary montone circuit that computes it. The *monotone complexity* of a monotone function is the smallest size of a binary montone circuit that computes it. Until 1985, the largest known lower bound for the monotone complexity of a monotone NP function of n variables was

$4n$. This was considerably improved in the fundamental paper of Razborov (1985), where a bound of $n^{\Omega(\log n)}$ to the clique$_k$-function (which is 1 iff a given graph contains a clique of size k) is established. Shortly afterward, Andreev (1985) used similar methods to obtain an exponential lower bound to a somewhat unnatural NP-function. Alon and Boppana (1987) strengthened the combinatorial arguments of Razborov and proved an exponential lower bound for the monotone circuit complexity of the clique function. In this section we describe a special case of this bound by showing that there are no linear-size monotone circuits that decide if a given graph contains a triangle. Although this result is much weaker than the ones stated above, it illustrates nicely all the probabilistic considerations in the more complicated proofs and avoids some of the combinatorial subtleties, whose detailed proofs can be found in the above-mentioned papers.

Put $n = \binom{m}{2}$, and let $x_1, x_2, \ldots, x_n$ be n Boolean variables representing the edges of a graph on the set of vertices $\{1, 2, \ldots, m\}$. Let $T = T(x_1, \ldots, x_n)$ be the monotone Boolean function whose value is 1 if the corresponding graph contains a triangle. Clearly, there is a binary monotone circuit of size $O(m^3)$ computing T. Thus the following theorem is tight, up to a polylogarithmic factor.

Theorem 4.1. *The monotone circuit complexity of T is at least $\Omega(m^3/\log^4 m)$.*

Before we present the proof of this theorem, we introduce some notation and prove a simple lemma. For any Boolean function $f = f(x_1, \ldots, x_n)$, define $A(f) = \{(x_1, \ldots, x_n) \in \{0,1\}^n : f(x_1, \ldots, x_n) = 1\}$. Obviously, $A(f \vee g) = A(f) \cup A(g)$ and $A(f \wedge g) = A(f) \cap A(g)$. Let C be a monotone circuit of size s computing the function $f = f(x_1, \ldots, x_n)$. Clearly, C supplies a monotone straight-line program of length s computing f, that is, a sequence of functions $x_1, x_2, \ldots, x_n, f_1, \ldots, f_s$, where $f_s = f$ and each f_i, for $1 \le i \le s$, is either an Or or an And of two of the previous functions. By applying the operation A, we obtain a sequence $A(C)$ of subsets of $(0,1)^n$: $A_{-n} = A_{x_n}, \ldots, A_{-1} = A_{x_1}, A_1, \ldots, A_s$, where $A_{x_i} = A(x_i)$, $A_s = A(f)$, and each A_i, for $1 \le i \le s$, is either a union or an intersection of two of the previous subsets. Let us replace the sequence $A(C)$ by an *approximating sequence* $M(C)$: $M_{-n} = M_{x_n} = A_{x_n}, \ldots, M_{-1} = M_{x_1} = A_{x_1}, M_1, \ldots, M_s$ defined by replacing the union and intersection operations in $A(C)$ by the approximating operations $\sqcup$ and $\sqcap$, respectively. The exact definition of these two operations will be given later, in such a way that for all admissible M and L the inclusions

$$M \sqcup L \supseteq M \cup L \qquad \text{and} \qquad M \sqcap L \subseteq M \cap L \tag{9}$$

will hold. Thus $M_{x_i} = A_{x_i}$ for all $1 \le i \le n$, and if for some $1 \le j \le s$ we have $A_j = A_\ell \cup A_k$, then $M_j = M_\ell \sqcup M_k$, whereas if $A_j = A_\ell \cap A_k$, then $M_j = M_\ell \sqcap M_k$. In the former case, put $\delta^j_\sqcup = M_j - (M_\ell \cup M_k)$ and $\delta^j_\sqcap = \phi$, and in the latter case, put $\delta^j_\sqcap = (M_\ell \cap M_k) - M_j$ and $\delta^j_\sqcup = \phi$.

Lemma 4.2. *For all members M_i of $M(C)$,*

$$A_i - \left(\bigcup_{j\le i}\delta_{\sqcap}^j\right) \subseteq M_i \subseteq A_i \cup \bigcup_{j\le i}\delta_{\sqcup}^j. \tag{10}$$

Proof. We apply induction on i. For $i < 0$, $M_i = A_i$ and thus (10) holds. Assuming (10) holds for all M_j with $j < i$, we prove it for i. If $A_i = A_\ell \cup A_k$, then, by the induction hypothesis

$$M_i = M_\ell \cup M_k \cup \delta_{\sqcup}^i \subseteq A_\ell \cup A_k \cup \bigcup_{j\le i}\delta_{\sqcup}^j = A_i \cup \bigcup_{j\le i}\delta_{\sqcup}^j$$

and

$$M_i = M_\ell \sqcup M_k \subseteq M_\ell \cup M_k \subseteq \left(A_\ell - \left(\bigcup_{j\le \ell}\delta_{\sqcap}^j\right)\right) \cup \left(A_k - \left(\bigcup_{j\le k}\delta_{\sqcap}^j\right)\right)$$

$$\supseteq A_i - \left(\bigcup_{j\le i}\delta_{\sqcap}^j\right),$$

as needed. If $A_i = A_\ell \cap A_k$, the proof is similar. ■

Lemma 4.2 holds for any choice of the operations $\sqcup$ and $\sqcap$ that satisfies (9). In order to prove Theorem 4.1, we define these operations as follows. Put $r = 100\log^2 m$. For any set R of at most r edges on $V = \{1,2,\dots,m\}$, let $\lceil R\rceil$ denote the set of all graphs on V containing at least one edge of R. In particular, $\lceil\phi\rceil$ is the empty set. We also let $\lceil *\rceil$ denote the set of all graphs. The elements of $M(C)$ will all have the form $\lceil R\rceil$ or $\lceil *\rceil$. Note that $A_{x_i} = M_{x_i}$ is simply the set $\lceil R\rceil$ where R is a singleton containing the appropriate single edge. For two sets R_1 and R_2 of at most r edges each, we define $\lceil R_1\rceil \sqcap \lceil R_2\rceil = \lceil R_1 \cap R_2\rceil$, $\lceil R_1\rceil \sqcap \lceil *\rceil = \lceil R_1\rceil$, and $\lceil *\rceil \sqcap \lceil *\rceil = \lceil *\rceil$. Similarly, if $|R_1 \cup R_2| \le r$, we define $\lceil R_1\rceil \sqcup \lceil R_2\rceil = \lceil R_1 \cup R_2\rceil$, whereas if $|R_1 \cup R_2| > r$, then $\lceil R_1\rceil \sqcup \lceil R_2\rceil = \lceil *\rceil$. Finally, $\lceil *\rceil \sqcup \lceil R_1\rceil = \lceil *\rceil \sqcup \lceil *\rceil = \lceil *\rceil$.

We now prove Theorem 4.1 by showing that there is no monotone circuit of size $s < \binom{m}{3}/2r^2$ computing the function T. Indeed, suppose this is false and let C be such a circuit. Let $M(C) = M_{x_n},\dots,M_{x_1},M_1,\dots,M_s$ be an approximating sequence of length s obtained from C as described above. By Lemma 4.2,

$$A(T) - \left(\bigcup_{j\le s}\delta_{\sqcap}^j\right) \subseteq M_s \subseteq A(T) \cup \bigcup_{j\le s}\delta_{\sqcup}^j. \tag{11}$$

We consider two possible cases.

Case 1. $M_s = \lceil R\rceil$, where $|R| \le r$.

Let us choose a random triangle Δ on $\{1,2,\ldots,m\}$. Clearly,

$$\Pr(\Delta \in M_s) \le \frac{r\cdot(m-2)}{\binom{m}{3}} < \frac{1}{2}.$$

Moreover, for each fixed j, $j \le s$,

$$\Pr(\Delta \in \delta_{\sqcap}^{j}) \le \frac{r^2}{\binom{m}{3}}.$$

This is because if $\delta_{\sqcap}^{j} \ne \phi$, then $\delta_{\sqcap}^{j} = (\lceil R_1 \rceil \cap \lceil R_2 \rceil) - \lceil R_1 \cap R_2 \rceil$ for some two sets of edges R_1, R_2, each of cardinality at most r. The only triangles in this difference are those containing an edge from R_1 and another edge from R_2 (and no edge of both). Since there are at most r^2 such triangles, the last inequality follows. Since $s < \binom{m}{3}/2r^2$, the last two inequalities imply that $\Pr(\Delta \notin M_s$ and $\Delta \notin \bigcup_{j\le s} \delta_{\sqcap}^{j}) > 0$ and thus there is such a triangle Δ. Since this triangle belongs to $A(T)$, this contradicts (11), showing that Case 1 is impossible. ■

Case 2. $M_s = \lceil * \rceil$.

Let B be a random spanning complete bipartite graph on $V = \{1,2,\ldots,m\}$ obtained by coloring each vertex in V randomly and independently by 0 or 1 and taking all edges connecting vertices with distinct colors. Since M_s is the set of all graphs, $B \in M_s$. Also $B \notin A(T)$, as it contains no triangle. We claim that for every fixed j, $j \le s$,

$$\Pr(B \in \delta_{\sqcup}^{j}) \le 2^{-\sqrt{r}/2} < \frac{1}{m^5}. \tag{12}$$

Indeed, if $\delta_{\sqcup}^{j} \ne \phi$, then $\delta_{\sqcup}^{j} = \lceil * \rceil - (\lceil R_1 \rceil \cup \lceil R_2 \rceil)$, where $|R_1 \cup R_2| > r$. Consider the graph whose set of edges is $R_1 \cup R_2$. Let d be its maximum degree. By Vizing's theorem, the set of its edges can be partitioned into at most $d+1$ matchings. Thus either $d > \sqrt{r}/2$ or the size of the maximum matching in this graph is at least $\sqrt{r}/2$. It follows that our graph contains a set of $k = \sqrt{r}/2$ edges $e_1,\ldots,e_k$ which form either a star or a matching. In each of these two cases, $\Pr(e_i \in B) = 1/2$ and these events are mutually independent. Hence

$$\Pr(B \notin \lceil R_1 \rceil \cup \lceil R_2 \rceil) \le 2^{-\sqrt{r}/2},$$

implying (12). Note that a similar estimate can be established without Vizing's theorem by observing that B does not belong to $(\lceil R_1 \rceil \cup \lceil R_2 \rceil)$ if and only if the vertices in any connected component of the graph whose edges are $R_1 \cup R_2$ belong to the same color class of B.

Since $s < \binom{m}{3}/2r^2 < m^5$, inequality (12) implies that there is a bipartite B such that $B \in M_s$, $B \notin A(T)$, and $B \notin \bigcup_{j\le s} \delta_{\sqcup}^{j}$. This contradicts (11), shows that Case 2 is impossible, and hence completes the proof of Theorem 4.1. ■

5. FORMULAE

Recall that a formula is a circuit in which every fanout is at most 1. Unlike in the case of circuits, there are known superlinear lower bounds for the minimum size of formulae computing various explicit NP functions over the full binary basis. For a Boolean function $f = f(x_1,\ldots,x_n)$, let us denote by $L(f)$ the minimum number of And and Or gates in a formula that uses And, Or, and Not gates and computes f. By De Morgan rules, we may assume that all Not gates appear in the first level of this formula. We conclude this chapter with a simple result of Subbotovskaya (1961), which implies that for the parity function $f = x_1 \oplus \cdots \oplus x_n$, $L(f) \geq \Omega(n^{3/2})$. This bound was improved later by Khrapchenko (1971) to $L(f) = n^2 - 1$. However, we present here only the weaker $\Omega(n^{3/2})$ lower bound, not only because it demonstrates, once more, the power of relatively simple probabilistic arguments, but also because a rather simple modification of this proof enabled Andreev (1988) to obtain an $\Omega(n^{5/2}/(\log n)^{O(1)})$ lower bound for $L(g)$ for another NP-function $g = g(x_1,\ldots,x_n)$. This is at present the largest known lower bound for the formula—complexity of an NP function of n variables over a complete basis.

The method of Subbotovskaya is based on random restrictions similar to the ones used in Section 2. The main lemma is the following.

Lemma 5.1. *Let $f = f(x_1,\ldots,x_n)$ be a nonatom Boolean function of n variables. Then there is an i, $1 \leq i \leq n$, and an $\epsilon \in \{0,1\}$ such that for the function $g = f(x_1,\ldots,x_{i-1},\epsilon,x_{i+1},\ldots,x_n)$ of $n-1$ variables obtained from f by substituting $x_i = \epsilon$, the following inequality holds:*

$$(L(g)+1) \leq \left(1-\frac{3}{2n}\right)(L(f)+1) \leq \left(1-\frac{1}{n}\right)^{3/2}(L(f)+1).$$

Proof. Fix a formula F computing f with $l = L(f)$ And and Or gates. F can be represented by a binary tree, each of whose $l+1$ leaves is labeled by an atom x_i or $\overline{x}_i$. Let us choose, randomly, a variable x_i, $1 \leq i \leq n$, according to a uniform distribution, and assign to it a random binary value $\epsilon \in \{0,1\}$. When we substitute the values ϵ and $1-\epsilon$ to x_i and $\overline{x}_i$, respectively, the number of leaves in F is reduced; the expected number of leaves omitted in this manner is $(l+1)/n$. However, further reduction may occur. Indeed, suppose a leaf is labeled x_i and it feeds, say, an And gate $x_i \wedge H$ in F. Observe that we may assume that the variable x_i does not appear in the subformula H, as otherwise F can be simplified by substituting $x_i = 1$ in H. If $x_i = \epsilon = 0$, then H can be deleted once we substitute the value for x_i, thus further decreasing the number of leaves. Since the behavior of this effect is similar for an Or gate (and also for $\overline{x}_i$ instead of x_i), it follows that the expected number of additional leaves omitted is at least $(l+1)/2n$. Hence the expected number of remaining leaves in the simplified formula is $(l+1)[1-3/2n]$, as claimed. ■

By repeatedly applying Lemma 5.1 we obtain the following.

Corollary 5.2. *If $f = f(x_1, \ldots, x_n)$ and $L(f) \leq (n/k)^{3/2} - 1$, then one can assign values to $n-k$ variables so that the resulting function g is an atom.*

Proof. Repeated application of Lemma 5.1 $n-k$ times yields a g with

$$(L(g)+1) \leq \prod_{i=k+1}^{n} \left(1 - \frac{1}{i}\right)^{3/2} (L(f)+1) = \left(\frac{k}{n}\right)^{3/2} (L(f)+1) \leq 1.$$

Hence g is either x_i or $\overline{x}_i$ for some i. ■

Corollary 5.3. *For the parity function $f = x_1 \oplus \cdots \oplus x_n$,*

$$L(f) > \left(\frac{n}{2}\right)^{3/2} - 1.$$

The Probabilistic Lens: Maximal Antichains

A family $\mathcal{F}$ of subsets of $\{1,\ldots,n\}$ is called an *antichain* if no set of $\mathcal{F}$ is contained in another.

Theorem. *Let $\mathcal{F}$ be an antichain. Then*

$$\sum_{A\in\mathcal{F}} \frac{1}{\binom{n}{|A|}} \leq 1.$$

Proof. Let σ be a uniformly chosen permutation of $\{1,\ldots,n\}$ and set

$$C_\sigma = \{\{\sigma(j) : 1 \leq j \leq i\} : 0 \leq i \leq n\}.$$

(The cases $i = 0, n$ give $\emptyset$, $\{1,\ldots,n\} \in C$, respectively.) Define a random variable

$$X = |\mathcal{F} \cap C_\sigma|.$$

We decompose

$$X = \sum_{A\in\mathcal{F}} X_A,$$

where X_A is the indicator random variable for $A \in C$. Then

$$E[X_A] = \Pr[A \in C_\sigma] = \frac{1}{\binom{n}{|A|}},$$

since C_σ contains precisely one set of size $|A|$, which is distributed uniformly among the $|A|$-sets. By linearity of expectation,

$$E[X] = \sum_{A\in\mathcal{F}} \frac{1}{\binom{n}{|A|}}.$$

For *any* σ, C_σ forms a chain—every pair of sets is comparable. Since $\mathcal{F}$ is an antichain, we *must* have $X = |\mathcal{F} \cap C_\sigma| \leq 1$. Thus $E[X] \leq 1$. ■

Corollary (Sperner's Theorem). *Let $\mathcal{F}$ be an antichain. Then*

$$|\mathcal{F}| \leq \binom{n}{\lfloor n/2 \rfloor}.$$

Proof. The function $\binom{n}{x}$ is maximized at $x = \lfloor n/2 \rfloor$, so that

$$1 \geq \sum_{A \in \mathcal{F}} \frac{1}{\binom{n}{|A|}} \geq \frac{|\mathcal{F}|}{\binom{n}{\lfloor n/2 \rfloor}}.$$

■

12

Discrepancy

1. BASICS

Suppose we are given a finite family of finite sets. Our object is to color the underlying points red and blue so that all of the sets have nearly the same number of red and blue points. It may be that our cause is hopeless—if the family consists of all subsets of a given set Ω, then regardless of the coloring, some set, either the red or the blue points, will have size at least half that of Ω and be monochromatic. In the other extreme, should the sets of the family be disjoint, then it is trivial to color so that all sets have the same number of red and blue points or, at worst, if the cardinality is odd, the number of red and blue points differing by only 1. The discrepancy will measure how good a coloring we may find.

To be formal, let a family $\mathcal{A}$ of subsets of Ω be given. Rather than using red and blue, we consider colorings as maps

$$\chi : \Omega \longrightarrow \{-1, +1\}.$$

For any $A \subset \Omega$, we set

$$\chi(A) = \sum_{a \in A} \chi(a).$$

Define the discrepancy of $\mathcal{A}$ with respect to χ by

$$\text{disc}(\mathcal{A}, \chi) = \max_{A \in \mathcal{A}} |\chi(A)|$$

and the discrepancy of $\mathcal{A}$ by

$$\text{disc}(\mathcal{A}) = \min_{\chi : \Omega \to \{-1,+1\}} \text{disc}(\mathcal{A}, \chi).$$

Other equivalent definitions of discrepancy reveal its geometric aspects. Let $\mathcal{A} = \{S_1, \ldots, S_m\}$, $\Omega = \{1, \ldots, n\}$, and let $B = [b_{ij}]$ be the m by n incidence matrix: $b_{ij} = 1$ if $j \in S_i$, otherwise $b_{ij} = 0$. A coloring χ may be associated with the vector $u = (\chi(1), \ldots, \chi(n)) \in \{-1, +1\}^n$ so that $Bu^T = (\chi(S_1), \ldots, \chi(S_m))$ and

$$\text{disc}(\mathcal{A}) = \min_{u \in \{-1,+1\}^n} |Bu^T|_\infty,$$

where $|v|_\infty$ is the L^∞-norm, the maximal absolute value of the coordinates. Similarly, letting v_j denote the jth column vector of B (the profile of point j),

$$\text{disc}(\mathcal{A}) = \min|\pm v_1 \pm \cdots \pm v_n|_\infty,$$

where the minimum ranges over all 2^n choices of sign.

We will generally be concerned with upper bounds to the discrepancy. Unraveling the definitions, $\text{disc}(\mathcal{A}) \leq K$ if and only if there *exists* a coloring χ for which $|\chi(A)| \leq K$ for all $A \in \mathcal{A}$. Naturally, we try the random coloring.

Theorem 1.1. *Let $\mathcal{A}$ be a family of n subsets of an m-set Ω. Then*

$$\text{disc}(\mathcal{A}) \leq \sqrt{2m\ln(2n)}.$$

Proof. Let $\chi : \Omega \longrightarrow \{-1,+1\}$ be random. For $A \subset \Omega$, let X_A be the indicator random variable for $|\chi(A)| > \alpha$ where we set $\alpha = \sqrt{2m\ln(2n)}$. If $|A| = a$, then $\chi(A)$ has distribution S_a, so by Theorem A.1 in Appendix A,

$$E[X_A] = \Pr[|\chi(A)| > \alpha] < 2e^{-\alpha^2/2a} \leq 2e^{-\alpha^2/2m} = \frac{1}{n}$$

by our propitious choice of α. Let X be the number of $A \in \mathcal{A}$ with $|\chi(A)| > \alpha$ so that

$$X = \sum_{A\in\mathcal{A}} X_A$$

and linearity of expectation gives

$$E[X] = \sum_{A\in\mathcal{A}} E[X_A] < |\mathcal{A}|\left(\frac{1}{n}\right) = 1.$$

Thus for some χ, we must have $X = 0$. This means $\text{disc}(\mathcal{A},\chi) \leq \alpha$ and therefore $\text{disc}(\mathcal{A}) \leq \alpha$. ■

2. SIX STANDARD DEVIATIONS SUFFICE

When $\mathcal{A}$ has both n sets and n points, Theorem 1.1 gives

$$\text{disc}(\mathcal{A}) = O\left(\sqrt{n\ln(n)}\right).$$

This is improved by the following result. Its proof resembles that of the main result of Beck (1981). The approach via entropy was suggested by R. Boppana.

Theorem 2.1 (Spencer [1985]). *Let $\mathcal{A}$ be a family of n subsets of an n-element set Ω. Then*

$$\text{disc}(\mathcal{A}) < 6\sqrt{n}.$$

With $\chi : \Omega \to \{-1, +1\}$ random, $A \in \mathcal{A}$, $\chi(A)$ has zero mean and variance at most $\sqrt{n}$. If $|\chi(A)| > 6\sqrt{n}$, then $\chi(A)$ is at least 6 standard deviations off the mean. The probability of this occurring is very small, but a fixed positive constant and the number of sets $A \in \mathcal{A}$ is going to infinity. In fact, a random χ almost always will *not* work. The specific constant 6 (actually 5.32) was the result of specific calculations that could certainly be further improved and will not concern us here. Rather, we show Theorem 2.1 with "6"= 11. A map

$$\chi : \Omega \longrightarrow \{-1, 0, +1\}$$

will be called a *partial* coloring. When $\chi(a) = 0$, we say a is uncolored. We define $\chi(A)$ as before.

Lemma 2.2. *Let $\mathcal{A}$ be a family of n subsets of an n-set Ω. Then there is a partial coloring χ with at most $10^{-9}n$ points uncolored such that*

$$|\chi(A)| \leq 10\sqrt{n}$$

for all $A \in \mathcal{A}$.

Here the values 10 and 10^{-9} are not best possible. The significant point is that they are absolute constants.

Proof of Lemma 2.2. Label the sets of $\mathcal{A}$ by $A_1, \ldots, A_n$ for convenience. Let

$$\chi : \Omega \longrightarrow \{-1, +1\}$$

be random. For $1 \leq i \leq n$, define

$$b_i = \text{nearest integer to } \frac{\chi(A_i)}{20\sqrt{n}}.$$

For example, $b_i = 0$ when $-10\sqrt{n} < \chi(A_i) < 10\sqrt{n}$ and $b_i = -3$ when $-70\sqrt{n} < \chi(A_i) < -50\sqrt{n}$. From Theorem A.1 (as in Theorem 1.1),

$$\Pr[b_i = 0] > 1 - 2e^{-50}$$

$$\Pr[b_i = 1] = \Pr[b_i = -1] < e^{-50}$$

$$\Pr[b_i = 2] = \Pr[b_i = -2] < e^{-450},$$

and, in general,

$$\Pr[b_i = s] = \Pr[b_i = -s] < e^{-50(2s-1)^2}.$$

Now we bound the *entropy* $H[b_i]$. Letting $p_j = \Pr[b_i = j]$,

$$\begin{aligned} H(b_i) &= \sum_{j=-\infty}^{+\infty} -p_j \log_2(p_j) \\ &\leq (1-2e^{-50})[-\log_2(1-2e^{-50})] \\ &\quad + 2e^{-50}[-\log_2 e^{-50}] \\ &\quad + 2e^{-450}[-\log_2 e^{-450}] + \cdots . \end{aligned}$$

The infinite sum clearly converges and is strongly dominated by the second term. Calculation gives

$$H(b_i) \leq \epsilon = 3 \times 10^{-20}.$$

Now consider the n-tuple $(b_1,\ldots,b_n)$. Of course, there may be correlation among the b_i. Indeed, if S_i and S_j are nearly equal, then b_i and b_j will usually be equal. But entropy is subadditive. Hence

$$H((b_1,\ldots,b_n)) \leq \sum_{i=1}^{n} H(b_i) \leq \epsilon n.$$

If a random variable Z assumes no value with probability greater than 2^{-t}, then $H(Z) \geq t$. In contrapositive form, there is a particular n-tuple $(s_1,\ldots,s_n)$ so that

$$\Pr[(b_1,\ldots,b_n) = (s_1,\ldots,s_n)] \geq 2^{-\epsilon n}.$$

Our probability space was composed of the 2^n possible colorings χ, all equally likely. Thus, shifting to counting, there is a set C' consisting of at least $2^{(1-\epsilon)n}$ colorings $\chi : \Omega \longrightarrow \{-1,+1\}$, all having the same value $(b_1,\ldots,b_n)$.

Let us think of the class C of all colorings $\chi : \Omega \longrightarrow \{-1,+1\}$ as the Hamming cube $\{-1,+1\}^n$ endowed with the Hamming metric

$$\rho(\chi,\chi') = |\{a : \chi(a) \neq \chi'(a)\}|.$$

Daniel Kleitman (1966a) has proven that if $\mathcal{D} \subset C$ and

$$|\mathcal{D}| \geq \sum_{i \leq r} \binom{n}{i}$$

with $r \leq n/2$, then $\mathcal{D}$ has diameter at least $2r$. That is, the set of a given size with minimal diameter is the ball. ($\mathcal{D}$ has diameter at least r trivially, which would suffice to prove Lemma 2.2 and Theorem 2.1 with weaker values for the constants.) In our case, we may take $r = \alpha n$ as long as $\alpha < 1/2$ and

$$2^{H(\alpha)} \leq 2^{1-\epsilon}.$$

Calculation gives that we may take $\alpha = (1/2)(1-10^{-9})$ with room to spare. (Taylor series gives

$$H\left(\frac{1}{2} - x\right) \sim 1 - \frac{2}{\ln 2} x^2$$

for x small.) Thus C' has diameter at least $n(1-10^{-9})$. Let $\chi_1,\chi_2 \in C'$ be at maximal distance. We set

$$\chi = \frac{\chi_1 - \chi_2}{2}.$$

χ is a partial coloring of Ω. $\chi(a) = 0$ if and only if $\chi_1(a) = \chi_2(a)$, which occurs for $n - \rho(\chi_1,\chi_2) \leq 10^{-9}n$ coordinates a. Finally, and crucially, for each $1 \leq i \leq n$, the colorings χ_1,χ_2 yield the same value b_i, which means that $\chi_1(A_i)$ and $\chi_2(A_i)$ lie on a common interval of length $20\sqrt{n}$. Thus

$$|\chi(A_i)| = \left|\frac{\chi_1(A_i) - \chi_2(A_i)}{2}\right| \leq 10\sqrt{n},$$

as desired. ■

Theorem 2.1 requires a coloring of all points, whereas Lemma 2.2 leaves $10^{-9}n$ points uncolored. The idea, now, is to iterate the procedure of Lemma 2.2, coloring all but, say, $10^{-18}n$ of the uncolored points on the second coloration. We cannot apply Lemma 2.2 directly, since we have an asymmetric situation with n sets and only $10^{-9}n$ points.

Lemma 2.3. *Let $\mathcal{A}$ be a family of n subsets of an r-set Ω with $r \leq 10^{-9}n$. Then there is a partial coloring χ of Ω with at most $10^{-40}r$ points uncolored, so that*

$$|\chi(A)| < 10\sqrt{r}\sqrt{\ln\left(\frac{n}{r}\right)}$$

for all $A \in \mathcal{A}$.

Proof. We outline the argument, which leaves room to spare. Let $A_1,\ldots, A_n$ denote the sets of $\mathcal{A}$. Let $\chi : \Omega \longrightarrow \{-1,+1\}$ be random. For $1 \leq i \leq n$, define

$$b_i = \text{nearest integer to } \frac{\chi(A_i)}{20\sqrt{r}\sqrt{\ln(n/r)}}.$$

Now the probability that $b_i = 1$ is less than $(r/n)^{50}$. The entropy $H(b_i)$ is dominated by this term and is less than

$$3\left(\frac{r}{n}\right)^{50}\left[-\log_2\left(\left(\frac{r}{n}\right)^{50}\right)\right] < 10^{-100}\frac{r}{n}.$$

The entropy of $(b_1,\ldots,b_r)$ is then less than $10^{-100}r$, one finds nearly antipodal χ_1,χ_2 with the same b's and takes $\chi = (\chi_1 - \chi_2)/2$ as before. ■

Proof of Theorem 2.1. Apply Lemma 2.2 to find a partial coloring χ^1 and then apply Lemma 2.3 repeatedly on the remaining uncolored points giving $\chi^2,\chi^3,\ldots$ until all points have been colored. Let χ denote the final coloring. For any $A \in \mathcal{A}$,

$$\chi(A) = \chi^1(A) + \chi^2(A) + \cdots$$

so that

$$|\chi(\mathcal{A})| \le 10\sqrt{n} + 10\sqrt{10^{-9}n}\sqrt{\ln 10^9} + 10\sqrt{10^{-49}n}\sqrt{\ln 10^{49}} + 10\sqrt{10^{-89}n}\sqrt{\ln 10^{89}} + \cdots$$

Removing the common $\sqrt{n}$ term gives a clearly convergent infinite series, strongly dominated by the first term, so that

$$|\chi(A)| \le 11\sqrt{n}$$

with room to spare. ■

Suppose that $\mathcal{A}$ consists of n sets on r points and $r < n$. We can apply Lemma 2.3 repeatedly (first applying Lemma 2.2 if $r > 10^{-9}n$) to give a coloring χ with

$$\mathrm{disc}(\mathcal{A}, \chi) < K\sqrt{r}\sqrt{\ln\left(\frac{n}{r}\right)},$$

where K is an absolute constant. As long as $r = n^{1-o(1)}$, this improves the random coloring result of Theorem 2.2.

3. LINEAR AND HEREDITARY DISCREPANCY

We now suppose that $\mathcal{A}$ has more points than sets. We write $\mathcal{A} = \{A_1, \ldots, A_n\}$ and $\Omega = \{1, \ldots, m\}$ and assume $m > n$. Note that $\mathrm{disc}(\mathcal{A}) \le K$ is equivalent to the existence of a set S, namely $S = \{j : \chi(j) = +1\}$, with $|S \cap A|$ within $K/2$ of $|A|/2$ for all $A \in \mathcal{A}$. We define the *linear discrepancy* of $\mathcal{A}$ by

$$\mathrm{lindisc}(\mathcal{A}) = \max_{p_1, \ldots, p_m \in [0,1]} \; \min_{\epsilon_1, \ldots, \epsilon_m \in \{0,1\}} \; \max_{A \in \mathcal{A}} \left| \sum_{i \in A} (\epsilon_i - p_i) \right| .$$

The upper bound $\mathrm{lindisc}(\mathcal{A}) \le K$ means that given any $p_1, \ldots, p_m$, there is a "simultaneous roundoff" $\epsilon_1, \ldots, \epsilon_m$ so that, with $S = \{j : \epsilon_j = 1\}$, $|S \cap A|$ is within K of the weighted sum $\sum_{j \in A} p_j$ for all $A \in \mathcal{A}$. Taking all $p_j = 1/2$, the upper bound implies $\mathrm{disc}(\mathcal{A}) \le 2K$. But $\mathrm{lindisc}(\mathcal{A}) \le K$ is much stronger. It implies, taking all $p_j = 1/3$, the existence of an S with all $|S \cap A|$ within K of $|A|/3$, and much more. Linear discrepancy and its companion, hereditary discrepancy, defined below, have been developed in Lovász, Spencer, Vesztergombi (1986). For $X \subset \Omega$, let $\mathcal{A}|_X$ denote the restriction of $\mathcal{A}$ to X, i.e., the family $\{A \cap X : A \in \mathcal{A}\}$. The next result "reduces" the bounding of $\mathrm{disc}(\mathcal{A})$ when there are more points than sets to the bounding of $\mathrm{lindisc}(\mathcal{A})$ when the points do not outnumber the sets.

Theorem 3.1. *Let $\mathcal{A}$ be a family of n sets on m points with $m \ge n$. Suppose that* $\mathrm{lindisc}(\mathcal{A}|_X) \le K$ *for every subset X of at most n points. Then*

$$\mathrm{lindisc}(\mathcal{A}) \le K.$$

Proof. Let $p_1, \ldots, p_m \in [0,1]$ be given. We define a reduction process. Call index j fixed if $p_j \in \{0,1\}$, otherwise call it floating, and let F denote the set of floating indices. If $|F| \leq n$, then halt. Otherwise, let y_j, $j \in F$, be a nonzero solution to the homogeneous system

$$\sum_{j \in A \cap F} y_j = 0, \qquad A \in \mathcal{A}.$$

Such a solution exists, since there are more variables ($|F|$) than equations (n), and may be found by standard techniques of linear algebra. Now set

$$p'_j = p_j + \lambda y_j, \qquad j \in F,$$

$$p'_j = p_j, \qquad j \notin F,$$

where we let λ be the real number of least absolute value so that, for some $j \in F$, the value p'_j becomes 0 or 1. Critically,

$$\sum_{j \in A} p'_j = \sum_{j \in A} p_j + \lambda \sum_{j \in A \cap F} y_j = \sum_{j \in A} p_j \qquad (*)$$

for all $A \in \mathcal{A}$. Now iterate this process with the new p'_j. At each iteration, at least one floating j becomes fixed and so the process eventually halts at some $p^*_1, \ldots, p^*_m$. Let X be the set of floating j at this point. Then $|X| \leq n$. By assumption, there exist ϵ_j, $j \in X$, so that

$$\left| \sum_{j \in A \cap X} p^*_j - \epsilon_j \right| \leq K, \qquad A \in \mathcal{A}.$$

For $j \notin X$, set $\epsilon_j = p^*_j$. As $(*)$ holds at each iteration,

$$\sum_{j \in A} p^*_j = \sum_{j \in A} p_j$$

and hence

$$\left| \sum_{j \in A} (p_j - \epsilon_j) \right| = \left| \sum_{j \in A} (p_j - p^*_j) + \sum_{j \in A \cap X} (p^*_j - \epsilon_j) \right| \leq K$$

for all $A \in \mathcal{A}$. ■

We now define the *hereditary discrepancy* of $\mathcal{A}$ by

$$\text{herdisc}(\mathcal{A}) = \max_{X \subseteq \Omega} \text{disc}(\mathcal{A}|_X).$$

Example. Let $\Omega = \{1, \ldots, n\}$ and let $\mathcal{A}$ consist of all intervals $[i,j] = \{i, i+1, \ldots, j\}$ with $1 \leq i \leq j \leq n$. Then $\text{disc}(\mathcal{A}) = 1$, as we may color Ω alternately $+1$ and -1. But also $\text{herdisc}(\mathcal{A}) = 1$. For given any $X \subseteq \Omega$, say with elements

$x_1 < x_2 < \cdots < x_r$, we may color X alternately by $\chi(x_k) = (-1)^k$. For any set $[i,j] \in \mathcal{A}$, the elements of $[i,j] \cap X$ are alternately colored. ■

Theorem 3.2. lindisc($\mathcal{A}$) $\leq$ herdisc($\mathcal{A}$).

Proof. Set $K =$ herdisc($\mathcal{A}$). Let $\mathcal{A}$ be defined on $\Omega = \{1,\ldots,m\}$ and let $p_1 \ldots, p_m \in [0,1]$ be given. First let us assume that all p_i have finite expansions when written in base 2. Let T be the minimal integer so that all $p_i 2^T \in Z$. Let J be the set of i for which p_i has a 1 in the Tth digit of its binary expansion, that is, so that $p_i 2^{T-1} \notin Z$. As disc($\mathcal{A}|_J$) $\leq K$, there exist $\epsilon_j \in \{-1,+1\}$, so that

$$\left| \sum_{j \in J \cap A} \epsilon_j \right| \leq K$$

for all $A \in \mathcal{A}$. Write $p_j = p_j^{(T)}$. Now set

$$p_j^{(T-1)} = \begin{cases} p_j^{(T)} & \text{if } j \notin J \\ p_j^{(T)} + \epsilon_j 2^{-T} & \text{if } j \in J. \end{cases}$$

That is, the $p_j^{(T-1)}$ are the "roundoffs" of the $p_j^{(T)}$ in the Tth place. Note that all $p_j^{(T-1)} 2^{-(T-1)} \in Z$. For any $A \in \mathcal{A}$,

$$\left| \sum_{j \in A} p_j^{(T-1)} - p_j^{(T)} \right| = \left| \sum_{j \in J \cap A} 2^{-T} \epsilon_j \right| \leq 2^{-T} K.$$

Iterate this procedure, finding $p_j^{(T-2)}, \ldots, p_j^{(1)}, p_j^{(0)}$. All $p_j^{(0)} 2^{-0} \in Z$, so all $p_j^{(0)} \in \{0,1\}$ and

$$\left| \sum_{j \in A} p_j^{(0)} - p_j^{(T)} \right| \leq \sum_{i=1}^{T} \left| \sum_{j \in A} p_j^{(i-1)} - p_j^{(i)} \right| \leq \sum_{i=1}^{T} 2^{-i} K \leq K,$$

as desired.

What about general $p_1, \ldots, p_m \in [0,1]$? We can be flip and say that, at least to a computer scientist, all real numbers have finite binary expansions. More rigorously, the function

$$f(p_1, \ldots, p_m) = \min_{\epsilon_1, \ldots, \epsilon_m \in \{0,1\}} \max_{A \in \mathcal{A}} \left| \sum_{i \in A} (\epsilon_i - p_i) \right|$$

is the finite minimum of finite maxima of continuous functions and thus is continuous. The set of $p_1, \ldots, p_m \in [0,1]$ with all $p_i 2^T \in Z$ for some T is a dense subset of $[0,1]$. As $f \leq K$ on this dense set, $f \leq K$ for all $p_1, \ldots, p_m \in [0,1]$. ■

Corollary 3.3. *Let $\mathcal{A}$ be a family of n sets on m points. Suppose* $\operatorname{disc}(\mathcal{A}|_X) \leq K$ *for every subset X with at most n points. Then* $\operatorname{disc}(\mathcal{A}) \leq 2K$.

Proof. For every $X \subseteq \Omega$ with $|X| \leq n$, $\operatorname{herdisc}(\mathcal{A}|_X) \leq K$, so by Theorem 3.2, $\operatorname{lindisc}(\mathcal{A}|_X) \leq K$. By Theorem 3.1, $\operatorname{lindisc}(\mathcal{A}) \leq K$. But

$$\operatorname{disc}(\mathcal{A}) \leq 2 \operatorname{lindisc}(\mathcal{A}) \leq 2K.$$ ∎

Corollary 3.4. *For any family $\mathcal{A}$ of n sets of arbitrary size,*

$$\operatorname{disc}(\mathcal{A}) \leq 12\sqrt{n}.$$

Proof. Apply Theorem 2.1 and Corollary 3.3. ∎

4. LOWER BOUNDS

We now give two quite different proofs that, up to a constant factor, Corollary 3.4 is best possible. A Hadamard matrix is a square matrix $H = (h_{ij})$ with all $h_{ij} \in \{-1, +1\}$ and with row vectors mutually orthogonal (and hence with column vectors mutually orthogonal). Let H be a Hadamard matrix of order n and let $v = (v_1, \ldots, v_n)$, $v_i \in \{-1, +1\}$. Then

$$Hv = v_1 c_1 + \cdots + v_n c_n,$$

where c_i denotes the ith column vector of H. Writing $Hv = (L_1, \ldots, L_n)$ and letting $|c|$ denote the usual Euclidean norm,

$$L_1^2 + \cdots + L_n^2 = |Hv|^2 = v_1^2 |c_1|^2 + \cdots + v_n^2 |c_n|^2 = n + \cdots + n = n^2,$$

since the c_i are mutually orthogonal. Hence some $L_i^2 \geq n$ and thus

$$|Hv|_\infty = \max(|L_1|, \ldots, |L_n|) \geq \sqrt{n}.$$

Now we transfer this result to one on families of sets. Let H be a Hadamard matrix of order n with first row and first column all ones. (Any Hadamard matrix can be so "normalized" by multiplying appropriate rows and columns by -1.) Let J denote the all-ones matrix of order n. Let $v_1, \ldots, L_1, \ldots$ be as above. Then

$$L_1 + \cdots + L_n = \sum_{i,j=1}^{n} v_j h_{ij} = \sum_{j=1}^{n} v_j \sum_{i=1}^{n} h_{ij} = n v_1 = \pm n,$$

since the first column sums to n but the other columns, being orthogonal to it, sum to 0. Set $\lambda = v_1 + \cdots + v_n$ so that $Jv = (\lambda, \ldots, \lambda)$ and

$$(H + J)v = (L_1 + \lambda, \ldots, L_n + \lambda).$$

We calculate

$$|(H+J)v|^2 = \sum_{i=1}^{n}(L_i+\lambda)^2 = \sum_{i=1}^{n}(L_i^2+2\lambda L_i+\lambda^2) = n^2 \pm 2n\lambda + n\lambda^2.$$

Assume n is even. (Hadamard matrices don't exist for odd n, except $n = 1$.) Then λ is an even integer. The quadratic (in λ) $n^2 \pm 2n\lambda + n\lambda^2$ has a minimum at ∓ 1, and so under the restiction of being an even integer, its minimum is at $\lambda = 0, \mp 2$, and so

$$|(H+J)v|^2 \geq n^2.$$

Again, some coordinate must be at least $\sqrt{n}$. Setting $H^* = (H+J)/2$,

$$|H^* v|_\infty \geq \frac{\sqrt{n}}{2}.$$

Let $\mathcal{A} = \{A_1, \ldots, A_m\}$ be any family of subsets of $\Omega = \{1, \ldots, n\}$ and let M denote the corresponding $m \times n$ incidence matrix. A coloring $\chi : \Omega \longrightarrow \{-1, +1\}$ corresponds to a vector $v = (\chi(1), \ldots, \chi(n)) \in \{-1, +1\}^n$. Then

$$\operatorname{disc}(\mathcal{A}, \chi) = |Mv|_\infty$$

and

$$\operatorname{disc}(\mathcal{A}) = \min_{v \in \{-1,+1\}^n} |Mv|_\infty.$$

In our case, H^* has entries 0, 1. Thus we have:

Theorem 4.1. *If a Hadamard matrix exists of order $n > 1$, then there exists a family $\mathcal{A}$ consisting of n subsets of an n-set with*

$$\operatorname{disc}(\mathcal{A}) \geq \frac{\sqrt{n}}{2}.$$

Remark. While it is not known precisely for which n a Hadamard matrix exists (the Hadamard conjecture is that they exist for $n = 1, 2$ and all multiples of 4; see, e.g., Hall [1986]), it is known that the orders of Hadamard matrices are dense in the sense that for all ϵ if n is sufficiently large, there will exist a Hadamard matrix of order between n and $n(1-\epsilon)$. This result suffices to extend Theorem 4.1 to an asymptotic result on all n.

Our second argument for the existence of $\mathcal{A}$ with high discrepancy involves turning the probabilistic argument "on its head." Let M be a random 0, 1 matrix of order n. Let $v = (v_1, \ldots, v_n), v_j = \pm 1$ be fixed and set $Mv = (L_1, \ldots, L_n)$. Suppose half of the $v_j = +1$ and half are -1. Then

$$L_i \sim B\left(\frac{n}{2}, \frac{1}{2}\right) - B\left(\frac{n}{2}, \frac{1}{2}\right),$$

which has roughly the normal distribution $N(0, \sqrt{n}/2)$. Pick $\lambda > 0$ so that

$$\int_{-\lambda}^{\lambda} \frac{1}{\sqrt{2\pi}} e^{-t^2/2}\, dt < \frac{1}{2}.$$

Then

$$\Pr\left[|L_i| < \frac{\lambda\sqrt{n}}{2}\right] < \frac{1}{2}.$$

When v is imbalanced, the same inequality holds; we omit the details. Now, crucially, the L_i are mutually independent, as each entry of M was independently chosen. Thus

$$\Pr\left[|L_i| < \frac{\lambda\sqrt{n}}{2} \text{ for all } 1 \le i \le n\right] < \left(\frac{1}{2}\right)^n.$$

There are "only" 2^n possible v. Thus the expected number of v for which $|Mv|_\infty < \lambda\sqrt{n}/2$ is less than $2^n 2^{-n} = 1$. For some M, this value must be 0; there are no such v. The corresponding family $\mathcal{A}$ thus has

$$\text{disc}(\mathcal{A}) > \frac{\lambda\sqrt{n}}{2}.$$

5. THE BECK–FIALA THEOREM

For any family $\mathcal{A}$, let $\deg(\mathcal{A})$ denote the maximal number of sets containing any particular point. The following result, due to J. Beck and T. Fiala (1981), uses only methods from linear algebra and thus is technically outside the scope we have set for this book. We include it both for the sheer beauty of the proof and because the result itself is very much in the spirit of this chapter.

Theorem 5.1. *Let $\mathcal{A}$ be a finite family of finite sets, with no restriction on either the number of sets nor on the cardinality of the sets, and* $\deg(\mathcal{A}) \le t$. *Then*

$$\text{disc}(\mathcal{A}) \le 2t - 1.$$

Proof. For convenience, write $\mathcal{A} = \{A_1, \ldots, A_m\}$, with all $A_i \subseteq \Omega = \{1, \ldots, n\}$. To each $j \in \Omega$, there is assigned a value x_j which will change as the proof progresses. Initially, all $x_j = 0$. At the end, all $x_j = \pm 1$. We will have $-1 \le x_j \le +1$ at all times and once $x_j = \pm 1$, it "sticks" there and that becomes its final value. A set S_i has value $\sum_{j \in S_i} x_j$. At any time, j is called fixed if $x_j = \pm 1$, otherwise it is *floating*. A set S_i is safe if it has fewer than t floating points, otherwise it is *active*. Note, crucially, that as points are in at most t sets and active sets contain more than t floating points, there must be fewer active sets than floating points.

We insist at all times that all active sets have value 0. This holds initially, since all sets have value 0. Suppose this condition holds at some stage. Consider x_j a variable for each floating j and a constant for each fixed j. The condition that S_i has value 0 then becomes a linear equation in these variables. This is an underdetermined system; there are fewer linear conditions (active sets) than variables (floating points). Hence we may find a line, parametrized

$$x'_j = x_j + \lambda y_j, \qquad j \text{ floating},$$

on which the active sets retain value 0. Let λ be the smallest value for which some x'_j becomes ± 1 and replace each x_j by x'_j. (Geometrically, follow the line until reaching the boundary of the cube in the space over the floating variables.) This process has left fixed variables fixed and so safe sets stayed safe sets (though active sets may have become safe), and so the condition still holds. In addition, at least one previously floating j has become fixed.

We iterate the above procedure until all j have become fixed. (Toward the end we may have no active sets, at which time we may simply set the floating x_j to ± 1 arbitrarily.) Now consider any set S_i. Initially, it had value 0 and it retained value 0 while it contained at least t floating points. Consider the time when it first becomes safe; say $1, \ldots, l$ were its floating points. At this moment its value is 0. The variables $y_1, \ldots, y_l$ can now change less than 2 to their final value, since all values are in $[-1, +1]$. Thus, in total, they may change less than $2t$. Hence the final value of S_i is less than $2t$ and, as it is an integer, it is at most $2t - 1$. ■

Conjecture. *If* $\deg(\mathcal{A}) \leq t$, *then* $\operatorname{disc}(\mathcal{A}) \leq K\sqrt{t}$, *$K$ an absolute constant.*

This conjecture seems to call for a melding of probabilistic methods and linear algebra. The constructions of t sets on t points, described in Section 4, show that, if true, this conjecture would be best possible.

The Probabilistic Lens: Unbalancing Lights

For any $m \times n$ matrix $B = (b_{ij})$ with coefficients $b_{ij} = \pm 1$, set

$$F[B] = \max_{x_i, y_j = \pm 1} \sum_{i=1}^{m} \sum_{j=1}^{n} x_i y_j b_{ij}.$$

As in Chapter 2, Section 5, we may interpret B as an m by n array of lights, each either on ($b_{ij} = +1$) or off ($b_{ij} = -1$). For each row and each column, there is a switch which, when pulled, changes all lights in that line from on to off or from off to on. Then $F[B]$ gives the maximal achievable number of lights on minus lights off. In the aforementioned section, we found a lower bound for $F[B]$ when $m = n$. Here we set $n = 2^m$ and find the precise best possible lower bound.

With $n = 2^m$ let A be an m by n matrix with coefficients ± 1 containing every possible column vector precisely once. We claim $F[A]$ is the minimal value of $F[B]$ over all m by n matrices B.

For any given B, let $x_1, \ldots, x_m = \pm 1$ be independently and uniformly chosen and set

$$X_j = \sum_{i=1}^{m} x_i b_{ij}, \qquad X = |X_1| + \cdots + |X_m|,$$

so that

$$F[B] = \max_{y_j = \pm 1} \max_{x_i = \pm 1} \sum_{j=1}^{n} y_j X_j = \max_{x_i = \pm 1} \sum_{j=1}^{n} |X_j| = \max X.$$

Regardless of the b_{ij}, X_i has distribution S_m, so that $E[|X_i|] = E[|S_m|]$ and, by linearity of expectation,

$$E[X] = nE[|S_m|].$$

With $B = A$, any choices of $x_1, \ldots, x_m = \pm 1$ have the effect of permuting the columns—the matrix $(x_i a_{ij})$ also has every column vector precisely once—so that $X = |X_1| + \cdots + |X_m|$ is a constant. Note that $E[X]$ is independent of B. In general, fixing $E[X] = \mu$, the minimal possible value for $\max X$ is achieved when X is the constant μ. Thus $F[B]$ is minimized with $B = A$.

13

Geometry

Suppose we choose randomly n points $P_1, \ldots, P_n$ on the unit circle, according to a uniform distribution. What is the probability that the origin lies in the convex hull of these points? There is a surprisingly simple (yet clever) way to compute this probability. Let us first choose n random pairs of antipodal points $Q_1, Q_{n+1} = -Q_1$, $Q_2, Q_{n+2} = -Q_2, \ldots, Q_n, Q_{2n} = -Q_n$ according to a uniform distribution. Notice that with probability 1, these pairs are all distinct. Next we choose each P_i to be either Q_i or its antipodal $Q_{n+i} = -Q_i$, where each choice is equally likely. Clearly, this corresponds to a random choice of the points P_i. The probability that the origin does *not* belong to the convex hull of the points P_i, given the (distinct) points Q_j, is precisely $x/2^n$, where x is the number of subsets of the points Q_j contained in an open half-plane determined by a line through the origin, which does not pass through any of the points Q_j. It is easy to see that $x = 2n$. This is because if we renumber the points Q_j so that their cyclic order on the circle is $Q_1, \ldots, Q_n, Q_{n+1}, \ldots, Q_{2n}$, and $Q_{n+i} = -Q_i$, then the subsets contained in such half-planes are precisely $\{Q_i, \ldots, Q_{n+i-1}\}$, where the indices are reduced modulo $2n$. Therefore the probability that the origin is in the convex hull of n randomly chosen points on the unit circle is precisely $1 - 2n/2^n$. Observe that the same result holds if we replace the unit circle by any centrally symmetric bounded planar domain with center 0 and that the argument can be easily generalized to higher dimensions.

This result, due to Wendel (1962), shows how in some cases a clever idea can replace a tedious computation. It also demonstrates the connection between probability and geometry. The probabilistic method has been recently used extensively for deriving results in discrete and computational geometry. Some of these results are described in this chapter.

1. THE GREATEST ANGLE AMONG POINTS IN EUCLIDEAN SPACES

There are several striking examples, in different areas of combinatorics, where the probabilistic method supplies very simple counterexamples to long-standing conjectures. Here is an example, due to Erdős and Füredi (1983).

Theorem 1.1. *For every $d \geq 1$, there is a set of at least $\lfloor \frac{1}{2}(2/\sqrt{3})^d \rfloor$ points in the d-dimensional Euclidean space R^d, such that all angles determined by three points from the set are strictly less than $\pi/2$.*

This theorem disproves an old conjecture of Danzer and Grünbaum (1962), that the maximum cardinality of such a set is at most $2d-1$. We note that, as proved by Danzer and Grünbaum, the maximum cardinality of a set of points in R^d in which all angles are at most $\pi/2$ is 2^d.

Proof. We select the points of a set X in R^d from the vertices of the d-dimensional cube. As usual, we view the vertices of the cube, which are 0,1-vectors of length d, as the characteristic vectors of subsets of a d-element set; i.e., each 0,1-vector a of length d is associated with the set

$$A = \{i : 1 \leq i \leq d,\ a_i = 1\}.$$

A simple consequence of Pythagoras's theorem gives that the three vertices a, b, and c of the d-cube, corresponding to the sets A, B, and C, respectively, determine a right angle at c if and only if

$$A \cap B \subset C \subset A \cup B. \tag{1}$$

As the angles determined by triples of points of the d-cube are always at most $\pi/2$, it suffices to construct a set X of cardinality, at least the one stated in the theorem, no three distinct members of which satisfy (1).

Define $m = \lfloor 1/2(2/\sqrt{3})^d \rfloor$, and choose, randomly and independently, $2m$ d-dimensional 0,1-vectors $a_1, \ldots, a_{2m}$, where each coordinate of each of the vectors independently is chosen to be either 0 or 1 with equal probability. For every fixed triple a, b, and c of the chosen points, the probability that the corresponding sets satisfy equation (1) is precisely $(3/4)^d$. This is because (1) simply means that for each i, $1 \leq i \leq d$, neither $a_i = b_i = 0$, $c_i = 1$, nor $a_i = b_i = 1$, $c_i = 0$, hold. Therefore the probability that for three fixed indices i, j, and k, our chosen points, a_i, a_j, a_k, form a right angle at a_k is $(3/4)^d$. Since there are $\binom{2m}{3}3$ possible triples that can produce such angles, the expected number of right angles is

$$\binom{2m}{3} 3 \left(\frac{3}{4}\right)^d \leq m,$$

where the last inequality follows from the choice of m. Thus there is a choice of a set X of $2m$ points in which the number of right angles is at most m. By deleting 1 point from each such angle, we obtain a set of at least $2m - m = m$ points in which all angles are strictly less than $\pi/2$. Notice that the remaining points are all distinct, since (1) is trivially satisfied if $A = C$. This completes the proof. ∎

It is worth noting that, as observed by Erdős and Füredi (1983), the proof above can be easily modified to give the following.

Theorem 1.2. *For every $\epsilon > 0$, there is a $\delta > 0$ such that for every $d \geq 1$, there is a set of at least $(1+\delta)^d$ points in R^d so that all the angles determined by three distinct points from the set are at most $\pi/3 + \epsilon$.*

We omit the detailed proof of this result.

2. EMPTY TRIANGLES DETERMINED BY POINTS IN THE PLANE

For a finite set X of points in general position in the plane, let $f(X)$ denote the number of *empty* triangles determined by triples of points of X, i.e., the number of triangles determined by points of X that contain no other point of X. Katchalski and Meir (1988) studied the minimum possible value of $f(X)$ for a set X of n points. Define $f(n) = \min\{f(X)\}$, as X ranges over all planar sets of n points in general position (i.e., containing no three collinear points). They proved that

$$\binom{n-1}{2} \leq f(n) < 200n^2.$$

These bounds were improved by Bárány and Füredi (1987), who showed that as n grows,

$$(1+o(1))n^2 \leq f(n) \leq (1+o(1))2n^2.$$

The construction that establishes the upper bound is probabilistic, and is given in the follwing theorem.

Theorem 2.1. *Let $I_1, I_2, \ldots, I_n$ be parallel unit intervals in the plane, where*

$$I_i = \{(x,y) : x = i,\ 0 \leq y \leq 1\}.$$

For each i, let us choose a point p_i randomly and independently from I_i, according to a uniform distribution. Let X be the set consisting of these n randomly chosen points. Then the expected number of empty triangles in X is at most $2n^2 + O(n \log n)$.

Clearly, with probability 1, X is a set of points in general position and hence the above theorem shows that $f(n) \leq 2n^2 + O(n \log n)$.

Proof. We first estimate the probability that the triangle determined by the points p_i, p_{i+a}, and p_{i+k} is empty, for some fixed i, a and $k = a + b \geq 3$. Let $A = (i,x)$, $B = (i+a,y)$, and $C = (i+k,z)$ be the points p_i, p_{i+a}, and p_{i+k}, respectively. Let m be the distance between B and the intersection point of the segment AC with the interval I_{i+a}. Since each of the points p_j for $i < j < i+k$ are chosen randomly according to a uniform distribution on I_j, it follows that the probability that the triangle determined by A, B and C is empty is precisely

$$\left(1-\frac{m}{a}\right)\left(1-2\frac{m}{a}\right)\cdots\left(1-(a-1)\frac{m}{a}\right)\left(1-(b-1)\frac{m}{b}\right)\cdots\left(1-\frac{m}{b}\right)$$
$$\leq \exp\left(-\frac{m}{a}-2\frac{m}{a}\cdots-(a-1)\frac{m}{a}-(b-1)\frac{m}{b}\cdots-\frac{m}{b}\right)$$
$$=\exp\left(-\binom{a}{2}\frac{m}{a}-\binom{b}{2}\frac{m}{b}\right)=\exp\left(-(k-2)\frac{m}{2}\right).$$

For every fixed choice of A and C, when the point $p_{i+a}=B$ is chosen randomly, the probability that its distance m from the intersection of the segment AC with the interval I_{i+a} is at most d is clearly at most $2d$, for all $d\geq 0$. Therefore the probability that the triangle determined by p_i, p_{i+a}, and p_{i+k} is empty is at most

$$2\int_{m\geq 0}\exp\left(-(k-2)\frac{m}{2}\right)dm=\frac{4}{k-2}.$$

It follows that the expected value of the total number of empty triangles is at most

$$n-2+\sum_{1\leq i\leq n-3}\ \sum_{3\leq k\leq n-i}\ \sum_{1\leq a\leq k-1}\frac{4}{k-2}$$
$$=n-2+\sum_{3\leq k\leq n-1}(n-k)\frac{4(k-1)}{k-2}$$
$$=n-2+\sum_{3\leq k\leq n-1}(n-k)\frac{4}{k-2}+4\sum_{3\leq k\leq n-1}(n-k)$$
$$=2n^2+O(n\log n).$$

This completes the proof. ■

The result above can be extended to higher dimensions by applying a similar probabilistic construction. A set X of n points in the d-dimensional Euclidean space is called *independent* if no $d+1$ of the points lie on a hyperplane. A simplex determined by $d+1$ of the points is called *empty* if it contains no other point of X. Let $f_d(X)$ denote the number of empty simplices of X, and define $f_d(n)=\min f_d(X)$, as X ranges over all independent sets of n points in R^d. Katchalski and Meir (1988) showed that $f_d(n)\geq\binom{n-1}{d}$. The following theorem of Bárány and Füredi (1987) shows that here again, a probabilistic construction gives a matching upper bound, up to a constant factor (which depends on the dimension). We omit the detailed proof.

Theorem 2.2. *There exists a constant $K=K(d)$, such that for every convex, bounded set $A\subset R^d$ with nonempty interior, if X is a random set of n points* *obtained by n random and independent choices of points of A picked with uniform distribution, then the expected number of empty simplices of X is at most $K\binom{n}{d}$.*

3. GEOMETRICAL REALIZATIONS OF SIGN MATRICES

Let $A = (a_{i,j})$ be an m by n matrix with $\{+1,-1\}$-entries. We say that A is *realizable* in R^d if there are m hyperplanes $H_1,\ldots,H_m$ in R^d passing through the origin and n points $P_1,\ldots,P_n$ in R^d, so that for all i and j, P_j lies in the positive side of H_i if $a_{i,j} = +1$, and in the negative side if $a_{i,j} = -1$. Let $d(A)$ denote the minimum dimension d such that A is realizable in R^d, and define $d(m,n) = \max(d(A))$, as A ranges over all m by n matrices with $+1,-1$-entries. Since $d(m,n) = d(n,m)$ we can consider only the case $m \geq n$.

The problem of determining or estimating $d(m,n)$, and in particular $d(n,n)$, was raised by Paturi and Simon (1984). This problem was motivated by an attempt to estimate the maximum possible "unbounded-error probabilistic communication complexity" of Boolean functions. Alon, Frankl, and Rödl (1985) proved that as n grows $n/32 \leq d(n,n) \leq (\frac{1}{2} + o(1))n$. Both the upper and the lower bounds are proved by combining probabilistic arguments with certain other ideas. In the next theorem, we prove the upper bound, which is probably closer to the truth.

Theorem 3.1. *For all $m \geq n$,*

$$d(m,n) \leq \frac{(n+1)}{2} + \sqrt{\frac{n-1}{2}\log m}.$$

For the proof, we need a definition and two lemmas. For a vector $\mathbf{a} = (a_1,\ldots, a_n)$ of $\{+1,-1\}$-entries, the number of *sign-changes* in $\mathbf{a}$ is the number of indices i, $1 \leq i \leq n-1$, such that $a_i = -a_{i+1}$. For a matrix A of $+1,-1$-entries, denote by $s(A)$ the maximum number of sign-changes in a row of A.

Lemma 3.2. *For any matrix A of $+1,-1$-entries, $d(A) \leq s(A) + 1$.*

Proof of Lemma 3.2. Let $A = (a_{i,j})$ be an m by n matrix of $+1,-1$ entries and suppose $s = s(A)$. Let $t_1 < t_2 < \cdots < t_n$ be arbitrary reals, and define n points $P_1,P_2,\ldots,P_n$ in R^{s+1} by $P_j = (1,t_j,t_j^2,\ldots,t_j^s)$. These points, whose last s coordinates represent points on the d-dimensional moment-curve, will be the points used in the realization of A. To complete the proof, we have to show that each row of A can be realized by a suitable hyperplane through the origin. This is proved by applying some of the known properties of the moment-curve as follows. Consider the sign-vector representing an arbitrary row of A. Suppose this vector has r sign-changes, where, of course, $r \leq s$. Suppose the sign-changes in this vector occur between the coordinates i_j and

i_j+1, for $1 \le j \le r$. Choose arbitrary reals $y_1, \ldots, y_r$, where $t_{i_j} < y_j < t_{i_j+1}$ for $1 \le j \le r$. Consider the polynomial $P(t) = \prod_{j=1}^{r}(t - y_j)$. Since its degree is at most s, there are real numbers a_j such that $P(t) = \sum_{j=0}^{s} a_j t^j$. Let H be the hyperplane in R^{s+1} defined by $H = \{(x_0, x_1, \ldots, x_s) \in R^{s+1} : \sum_{j=0}^{s} a_j x_j = 0\}$. Clearly, the point $P_j = (1, t_j, \ldots, t_j^s)$ is on the positive side of this hyperplane if $P(t_j) > 0$, and is on its negative side if $P(t_j) < 0$. Since the polynomial P changes sign only in the values y_j, it follows that the hyperplane H separates the points $P_1, \ldots, P_n$ according to the sign pattern of the corresponding row of A. Hence, by choosing the orientation of H appropriately, we conclude that A is realizable in R^{s+1}, completing the proof of the lemma. ■

Lemma 3.3. *For every m by n matrix A of $+1, -1$-entries, there is a matrix B obtained from A by multiplying some of the columns of A by -1, such that*

$$s(B) \le \frac{(n-1)}{2} + \sqrt{\frac{n-1}{2}\log m}.$$

Proof. For each column of A, randomly and independently, choose a number $\epsilon \in \{+1, -1\}$, where each of the two choices is equally likely, and multiply this column by ϵ. Let B be the random sign-matrix obtained in this way. Consider an arbitrary fixed row of B. One can easily check that the random variable describing the number of sign-changes in this row is a binomial random variable with parameters $n-1$ and $p = 1/2$. This is because no matter what the entries of A in this row are, the row of B is a totally random row of $-1, 1$-entries. By the standard estimates for binomial distributions, described in Appendix A, the probability that this number is greater than $(n-1)/2 + \sqrt{[(n-1)/2]\log m}$ is smaller than $1/m$. Therefore, with positive probability, the number of sign changes in each of the m rows is at most that large, completing the proof. ■

Proof of Theorem 3.1. Let A be an arbitrary m by n matrix of $+1, -1$-entries. By Lemma 3.3, there is a matrix B obtained from A by replacing some of its columns by their inverses, such that

$$s(B) \le \frac{(n-1)}{2} + \sqrt{\frac{n-1}{2}\log m}.$$

Observe that $d(A) = d(B)$, since any realization of one of these matrices by points and hyperplanes through the origin gives a realization of the other one by replacing the points corresponding to the altered columns by their antipodal points. Therefore, by Lemma 3.2,

$$d(A) = d(B) \le s(B) + 1 \le \frac{(n+1)}{2} + \sqrt{\frac{n-1}{2}\log m}.$$

This completes the proof. ■

It is worth noting that by applying the (general) six standard deviations theorem stated in the end of Chapter 12, Section 2, the estimate in Lemma 3.3 (and hence in Theorem 3.1) can be improved to $n/2 + O(\sqrt{n}\log(m/n))$. It can be also shown that if n and m grow so that m/n^2 tends to infinity and $(\log_2 m)/n$ tends to 0, then for almost all m by n matrices A of $+1, -1$-entries, $d(A) = (1/2 + o(1))n$.

4. ϵ-NETS AND VC-DIMENSIONS OF RANGE SPACES

What is the minimum number $f = f(n,\epsilon)$ such that every set X of n points in the plane contains a subset S of at most f points such that every triangle containing at least ϵn points of X contains at least 1 point of S? As we shall see in this section, there is an absolute constant c such that $f(n,\epsilon) \leq c/\epsilon \log 1/\epsilon$, and this estimate holds for every n. This somewhat surprising result is a very special case of a general theorem of Vapnik and Chervonenkis (1971), which has been extended by Haussler and Welzl (1987), and which has many interesting applications in computational geometry and in statistics. In order to describe this result, we need a few definitions. A *range space* S is a pair (X,R), where X is a (finite or infinite) set and R is a (finite or infinite) family of subsets of X. The members of X are called *points* and those of R are called *ranges*. If A is a subset of X, then $P_R(A) = \{r \cap A : r \in R\}$ is the *projection* of R on A. In case this projection contains all subsets of A, we say that A is *shattered*. The *Vapnik–Chervonenkis dimension* (or VC-*dimension*) of S, denoted by VC(S), is the maximum cardinality of a shattered subset of X. If there are arbitrarily large shattered subsets, then VC(S) $= \infty$.

The number of ranges in any finite range space with a given number of points and a given VC-dimension cannot be too large. For integers $n \geq 0$ and $d \geq 0$, define a function $g(d,n)$ by

$$g(d,n) = \sum_{i=0}^{d} \binom{n}{i}.$$

Observe that for all $n,d \geq 1$, $g(d,n) = g(d,n-1) + g(d-1,n-1)$. The following combinatorial lemma was proved, independently, by Sauer (1972), Perles and Shelah, and, in a slightly weaker form by Vapnik and Chervonenkis.

Lemma 4.1. *If (X,R) is a range space of* VC-*dimension d with $|X| = n$ points, then $|R| \leq g(d,n)$.*

Proof. We apply induction on $n + d$. The assertion is trivially true for $d = 0$ and $n = 0$. Assuming it holds for n and $d - 1$ and for $n - 1$ and $d - 1$, we prove it for n and d. Let $S = (X,R)$ be a range space of VC-dimension d on n points. Suppose $x \in X$, and consider the two range spaces $S - x$ and $S \backslash x$ defined as follows. $S - x = (X - \{x\}, R - x)$, where $R - x = \{r - \{x\} : r \in R\}$. $S \backslash x = (X - \{x\}, R \backslash x)$, where $R \backslash x = \{r \in R : x \notin r, r \cup \{x\} \in R\}$. Clearly, the

VC-dimension of $S-x$ is at most d. It is also easy to see that the VC-dimension of $S\backslash x$ is at most $d-1$. Therefore, by the induction hypothesis,

$$|R| = |R-x| + |R\backslash x| \leq g(d,n-1) + g(d-1,n-1) = g(d,n),$$

completing the proof. ■

It is easy to check that the estimate given in the above lemma is sharp for all possible values of n and d. If (X,R) is a range space of VC-dimension d and $A \subset X$, then the VC-dimension of $(A, P_R(A))$ is clearly at most d. Therefore the last lemma implies the following.

Corollary 4.2. *If (X,R) is a range space of VC-dimension d, then for every finite subset A of X, $|P_R(A)| \leq g(d,|A|)$.*

There are many range spaces with finite VC-dimension that arise naturally in discrete and computational geometry. One such example is the space $S = (R^d, H)$, whose points are all the points in the d-dimensional Euclidean space, and whose set of ranges is the set of all (open) half-spaces. Any set of $d+1$ affinely independent points is shattered in this space, and, by Radon's theorem, no set of $d+2$ points is shattered. Therefore $VC(S) = d+1$. As shown by Dudley (1978), if (X,R) has finite VC-dimension, so does (X,R_k), where R_k is the set of all Boolean combinations formed from at most k ranges in R. In particular, the following statement is a simple consequence of Corollary 4.2.

Corollary 4.3. *Let (X,R) be a range space of VC-dimension $d \geq 2$, and let (X,R_h) be the range space on X in which $R_h = \{(r_1 \cap \cdots \cap r_h) : r_1,\ldots,r_h \in R\}$. Then $VC(X,R_h) \leq 2dh\log(dh)$.*

Proof. Let A be an arbitrary subset of cardinality n of X. By Corollary 4.2, $|P_R(A)| \leq g(d,n) \leq n^d$. Since each member of $P_{R_h}(A)$ is an intersection of h members of $P_R(A)$, it follows that $|P_{R_h}(A)| \leq \binom{g(d,n)}{h} \leq n^{dh}$. Therefore, if $n^{dh} < 2^n$, then A cannot be shattered. But this inequality holds for $n \geq 2dh\log(dh)$, since $dh \geq 4$. ■

As shown above, the range space whose set of points is R^d, and whose set of ranges is the set of all half-spaces, has VC-dimension $d+1$. This and the last corollary imply that the range space (R^d, C_h), where C_h is the set of all convex d-polytopes with h facets, has a VC-dimension that does not exceed $2(d+1)h\log((d+1)h)$.

An interesting property of range spaces with a finite VC-dimension is the fact that each finite subset of such a set contains relatively small good samples in the sense described below. Let (X,R) be a range space and let A be a finite subset of X. For $0 \le \epsilon \le 1$, a subset $B \subset A$ is an *ϵ-sample* for A if for any range $r \in R$, the inequality

$$||A \cap r|/|A| - |B \cap r|/|B|| \le \epsilon$$

holds. Similarly, a subset $N \subset A$ is an *ϵ-net* for A if any range $r \in R$ satisfying $|r \cap A| > \epsilon A$ contains at least 1 point of N.

Notice that every ϵ-sample for A is also an ϵ-net, and that the converse is not true. However, both notions define subsets of A that represent approximately some of the behavior of A with respect to the ranges. Our objective is to show the existence of small ϵ-nets or ϵ-samples for finite sets in some range spaces. Observe that if (X,R) is a range space with an infinite VC-dimension, then for every n there is a shattered subset A of X of cardinality n. It is obvious that any ϵ-net (and hence certainly any ϵ-sample) for such an A must contain at least $(1-\epsilon)n$ points, i.e., it must contain almost all points of A. Therefore, in infinite VC-dimension, there are no small nets or samples. However, it turns out that in finite VC-dimension, there are always very small nets and samples. The following theorem was proved by Vapnik and Chervonenkis (1971).

Theorem 4.4. *There is a positive constant c such that if (X,R) is any range space of* VC*-dimension at most d, $A \subset X$ is a finite subset, and $\epsilon, \delta > 0$, then a random subset B of cardinality s of A where s is at least the minimum between $|A|$ and*

$$\frac{c}{\epsilon^2}\left(d \log \frac{d}{\epsilon} + \log \frac{1}{\delta}\right)$$

is an ϵ-sample for A with probability at least $1-\delta$.

Using similar ideas, Haussler and Welzl (1987) proved the following theorem.

Theorem 4.5. *Let (X,R) be a range space of* VC*-dimension d, let A be a finite subset of X, and suppose $0 < \epsilon$, $\delta < 1$. Let N be a set obtained by m random independent draws from A, where*

$$m \ge \max\left(\frac{4}{\epsilon}\log\frac{2}{\delta}, \frac{8d}{\epsilon}\log\frac{8d}{\epsilon}\right). \tag{2}$$

Then N is an ϵ-net for A with probability at least $1-\delta$.

Therefore, if A is a finite subset of a range space of finite VC-dimension d, then for any $\epsilon > 0$, A contains ϵ-nets as well as ϵ-samples whose size is at most some function of ϵ and d, independent of the cardinality of A! The result about the triangles mentioned in the first paragraph of this section thus follows from Theorem 4.5, together with the observation following Corollary 4.3 that implies that the range space whose ranges are all triangles in the plane has a finite VC-dimension. We note that, as shown by Pach and Woeginger (1990), there are cases in which for fixed δ, the dependence of m in $1/\epsilon$ cannot be linear, but there is no known natural geometric example demonstrating this phenomenon.

The proofs of Theorems 4.4 and 4.5 are very similar. Since the computation in the proof of Theorem 4.5 is simpler, we describe here only the proof of this theorem, and encourage the reader to try and make the required modifications that yield a proof for Theorem 4.4.

Proof of Theorem 4.4. Let (X,R) be a range space with VC-dimension d, and let A be a subset of X of cardinality $|A| = n$. Suppose m satisfies (2), and let $N = (x_1,\ldots,x_m)$ be obtained by m independent random choices of elements of A. (The elements in N are not necessarily distinct, of course). Let E_1 be the following event:

$$E_1 = \{\exists r \in R : |r \cap A| > \epsilon n, r \cap N = \emptyset\}.$$

To complete the proof, we must show that the probability of E_1 is at most δ. To this end, we make an additional random choice and define another event as follows. Independently of our previous choice, we let $T = (y_1,\ldots,y_m)$ be obtained by m independent random choices of elements of A. Let E_2 be the event defined by

$$E_2 = \left\{\exists r \in R : |r \cap A| > \epsilon n,\ r \cap N = \emptyset,\ |r \cap T| \geq \frac{\epsilon m}{2}\right\}.$$

(Since the elements of T are not necessarily distinct, the notation $|r \cap T|$ means here $|\{i : 1 \leq i \leq m,\ y_i \in r\}|$. The quantities $|r \cap N|$ and $|r \cap (N \cup T)|$ are similarly defined.) ■

Claim 4.6.

$$\Pr(E_2) \geq \tfrac{1}{2}\Pr(E_1).$$

Proof. It suffices to prove that the conditional probability $\Pr(E_2 \mid E_1)$ is at least $1/2$. Suppose that the event E_1 occurs. Then there is an $r \in R$ such that $|r \cap A| > \epsilon n$ and $r \cap N = \emptyset$. The conditional probability above is clearly at least the probability that for this specific r, $|r \cap T| \geq \epsilon m/2$. However, $|r \cap T|$ is a binomial random variable with expectation ϵm and variance $\epsilon(1-\epsilon)m \leq \epsilon m$, and hence, by Chebyschev's inequality

$$\Pr\left(|r \cap T| < \frac{\epsilon m}{2}\right) \leq \frac{\epsilon m}{(\epsilon m/2)^2} = \frac{4}{\epsilon m} \leq \frac{1}{2},$$

where the last inequality follows from (2). Thus, the assertion of Claim 4.6 is correct. ■

Claim 4.7.

$$Pr(E_2) \leq g(d,2m)2^{-(\epsilon m/2)}.$$

Proof. The random choice of N and T can be described in the following way, which is equivalent to the previous one. First one chooses $N \cup T = (z_1, \ldots, z_{2m})$ by making $2m$ random independent choices of elements of A, and then one chooses randomly precisely m of the elements z_i to be the set N (the remaining elements z_j form the set T, of course). For each range $r \in R$ satisfying $|r \cap A| > \epsilon n$, let E_r be the event that $|r \cap T| > \epsilon m/2$ and $r \cap N = \emptyset$. A crucial fact is that if $r, r' \in R$ are two ranges, $|r \cap A| > \epsilon n$ and $|r' \cap A| > \epsilon n$, and if $r \cap (N \cup T) = r' \cap (N \cup T)$, then the two events E_r and $E_{r'}$, when both are conditioned on the choice of $N \cup T$, are identical. This is because the occurrence of E_r depends only on the intersection $r \cap (N \cup T)$. Therefore, for any fixed choice of $N \cup T$, the number of distinct events E_r does not exceed the number of different sets in the projection $P_{N \cup T}(R)$. Since the VC-dimension of X is d, Corollary 4.2 implies that this number does not exceed $g(d, 2m)$.

Let us now estimate the probability of a fixed event of the form E_r, given the choice of $N \cup T$. This probability is at most

$$\Pr\left(r \cap N = \emptyset \mid |r \cap (N \cup T)| > \frac{\epsilon m}{2}\right).$$

Define $p = |r \cap (N \cup T)|$. Since the choice of N among the elements of $N \cup T$ is independent of the choice of $N \cup T$, the last conditional probability is precisely

$$\frac{(2m-p)(2m-p-1)\cdots(m-p+1)}{2m(2m-1)\cdots(m+1)}$$
$$= \frac{m(m-1)\cdots(m-p+1)}{2m(2m-1)\cdots(2m-p+1)} \leq 2^{-p} \leq 2^{-(\epsilon m/2)}.$$

Since there are at most $g(d, 2m)$ potential distinct events E_r, it follows that the probability that at least one of them occurs given the choice of $N \cup T$ is at most $g(d, 2m)2^{-(\epsilon m/2)}$. Since this estimate holds conditioned on every possible choice of $N \cup T$, it follows that the probability of the event E_2 is at most $g(d, 2m)2^{-(\epsilon m/2)}$. This establishes Claim 4.7.

By Claims 4.6 and 4.7, $\Pr(E_1) \leq 2g(d, 2m)2^{-(\epsilon m/2)}$. To complete the proof of the theorem, it remains to show that if m satisfies inequality (2), then

$$2g(d, 2m)2^{-(\epsilon m/2)} \leq \delta.$$

We describe the proof for $d \geq 2$. The computation for $d = 1$ is easier. Since $g(d, 2m) \leq (2m)^d$, it suffices to show that

$$2(2m)^d \leq \delta 2^{(\epsilon m/2)},$$

that is, that

$$\frac{\epsilon m}{2} \geq d \log(2m) + \log \frac{2}{\delta}.$$

From (2), it follows that

$$\frac{\epsilon m}{4} \geq \log \frac{2}{\delta},$$

and hence it suffices to show that

$$\frac{\epsilon m}{4} \geq d \log(2m).$$

The validity of the last inequality for some value of m implies its validity for any bigger m, and thus it suffices to check that it is satisfied for $m = 8d/\epsilon \log(8d/\epsilon)$, i.e., that

$$2d \log \frac{8d}{\epsilon} \geq d \log\left(\frac{16d}{\epsilon} \log \frac{8d}{\epsilon}\right).$$

The last inequality is equivalent to $4d/\epsilon \geq \log(8d/\epsilon)$, which is certainly true. This completes the proof of the theorem. ■

Theorems 4.4 and 4.5 have been used for constructing efficient data structures for various problems in computational geometry. A trivial example is just the observation that Theorem 4.4 implies the following: for every $\epsilon > 0$, there is a constant $c = c(\epsilon)$ such that for every n and every set A of n points in the plane, there is a data structure of size $c(\epsilon)$ that enables us to estimate, given any triangle in the plane, the number of points of A in this triangle up to an additive error of ϵn. This is done simply by storing the coordinates of a set of points that form an ϵ-sample for A considered as a subset of the range space whose ranges are all planar triangles. More sophisticated data structures whose construction relies on the above two theorems can be found in the paper of Haussler and Welzl (1987).

The Probabilistic Lens: Efficient Packing

Let $C \subset R^n$ be bounded with Riemann measure $\mu = \mu(C) > 0$. Let $N(C,x)$ denote the maximal number of disjoint translates of C that may be packed in a cube of side x and define the packing constant

$$\delta(C) = \frac{1}{\mu(C)} \lim_{x \to \infty} N(C,x)x^{-n},$$

the maximal proportion of space that may be packed by copies of C. The following result improves the one described in Chapter 3, Section 4.

Theorem. *Let C be bounded, convex, and centrally symmetric about the origin. Then*

$$\delta(C) \geq 2^{-(n-1)}.$$

Proof. Fix $\epsilon > 0$. Normalize so $\mu = \mu(C) = 2 - \epsilon$. For any real z, let C_z denote the "slab" of $(z_1, \ldots, z_{n-1}) \in R^{n-1}$ such that $(z_1, \ldots, z_{n-1}, z) \in C$ and let $\mu(C_z)$ be the usual $n-1$-dimensional measure of C_z. Riemann measurability implies

$$\lim_{\gamma \to 0} \sum_{m \in Z} \mu(C_{m\gamma})\gamma = \mu(C).$$

Let K be an integer sufficiently large so that

$$\sum_{m \in Z} \mu(C_{mK^{-(n-1)}})K^{-(n-1)} < 2$$

and further that all points of C have all coordinates less than $K/2$.

For $1 \leq i \leq n-1$, let $v_i \in R^n$ be that vector with all coordinates zero except K as the ith coordinate. Let

$$v = (z_1, \ldots, z_{n-1}, K^{-(n-1)}),$$

where $z_1, \ldots, z_{n-1}$ are chosen uniformly and independently from the real interval $[0, K)$. Let Λ_v denote the lattice generated by the v's—that is,

$$\Lambda_v = \{m_1 v_1 + \cdots + m_{n-1} v_{n-1} + mv : m_1, \ldots, m_{n-1}, m \in Z\}$$
$$= \{(mz_1 + m_1 K, \ldots, mz_{n-1} + m_{n-1} K, mK^{-(n-1)} : m_1, \ldots, m_{n-1}, m \in Z\}.$$

Let $\theta(x)$ denote that unique $x' \in (-K/2, K/2]$ so that $x - mK = x'$ for some $m \in Z$. For $m \in Z$, let A_m be the event that some $m_1 v_1 + \cdots + m_{n-1} v_{n-1} + mv \in C$. Since all coordinates of all points of C are less than $K/2$, A_m occurs if and only if

$$(\theta(mz_1), \ldots, \theta(mz_{n-1}), mK^{-(n-1)}) \in C,$$

which occurs if and only if $(\theta(mz_1), \ldots, \theta(mz_{n-1})) \in C_{mK^{-(n-1)}}$. The independence and uniformity of the z_i over $[0, K)$ imply the independence and uniformity of the $\theta(z_i)$ over $(-K/2, K/2]$, and so

$$\Pr[A_m] = K^{-(n-1)} \mu(C_{mK^{-(n-1)}}).$$

Summing over positive m, and employing the central symmetry,

$$\sum_{m>0} \Pr[A_m] < \frac{1}{2} \sum_{m \in Z} K^{-(n-1)} \mu(C_{mK^{-(n-1)}}) < \frac{1}{2} 2 = 1.$$

Hence there *exists* v with all A_m, $m > 0$, not holding. By the central symmetry, A_m and A_{-m} are the same event, so no A_m, $m \neq 0$, holds. When $m = 0$, the points $m_1 v_1 + \cdots + m_{n-1} v_{n-1} = K(m_1, \ldots, m_{n-1}, 0)$ all lie outside C except the origin. For this v,

$$\Lambda_v \cap C = \{0\}.$$

Consider the set of translates $C + 2w$, $w \in \Lambda_v$. Suppose

$$z = c_1 + 2w_1 = c_2 + 2w_2 \qquad \text{with} \quad c_1, c_2 \in C, \quad w_1, w_2 \in \Lambda_v.$$

Then $(c_1 - c_2)/2 = w_2 - w_1$. From convexity and central symmetry, $(c_1 - c_2)/2 \in C$. As $w_2 - w_1 \in \Lambda_v$, it is 0 and hence $c_1 = c_2$ and $w_1 = w_2$. That is, the translates form a packing of R^n. As $\det(2\Lambda_v) = 2^n \det(\Lambda_v) = 2^n$, this packing has density $2^{-n}\mu = 2^{-n}(2 - \epsilon)$. As $\epsilon > 0$ was arbitrary, $\delta(C) \geq 2^{-(n-1)}$. ■

14

Codes and Games

1. CODES

Suppose we want to send a message, here considered a string of bits, across a noisy channel. There is a probability p that any bit sent will be received incorrectly. The value p is a parameter of the channel and cannot be changed. We assume that p is both the probability that a sent 0 is received as a one and that a sent one is received as a 0. Sent bits are always received, but perhaps incorrectly. We further assume that the events that the bits are received incorrectly are mutually independent. The case $p = .1$ will provide a typical example.

How can we improve the reliability of the system? One simple way is to send each bit three times. When the three bits are received, we use majority rule to decode. The probability of incorrect decoding is then $3p^2(1-p)+p^3 = .028$ in our instance. We have sacrificed speed—the rate of transmission of this method is $1/3$—and gained accuracy in return. If we send each bit five times and use majority rule to decode, the probability of incorrect decoding drops to .00856 but the rate of transmission also drops to $1/5$. Clearly, we may make the probability of incorrect decoding as low as needed, but seemingly with the trade-off that the rate of transmission tends to 0. It is the fundamental theorem of information theory—due to Claude Shannon—that this trade-off is not necessary: There are codes with rate of transmission approaching a positive constant (dependent on p) with probability of incorrect transmission approaching 0.

A *coding scheme* consists of positive integers m, n, a function $f : \{0,1\}^m \to \{0,1\}^n$ called the encoding function, and a function $g : \{0,1\}^n \to \{0,1\}^m$ called the decoding function. The notion is that a message (or segment of message) $x \in \{0,1\}^m$ will be encoded and sent as $f(x)$ and a received message $y \in \{0,1\}^n$ will be decoded as $g(y)$. The rate of transmission of such a scheme is defined as m/n. Let $E = (e_1, \ldots, e_n)$ be a random string defined by $\Pr[e_i = 1] = p$, $\Pr[e_i = 0] = 1-p$, the values e_i mutually independent. We define the probability of correct transmission as $\Pr[g(f(x)+E) = x]$. Here x is assumed to be uniformly distributed over $\{0,1\}^m$ and independent of E, + is vector addition modulo 2.

A crucial role is played by the *entropy function*

$$H(p) = -p\log_2 p - (1-p)\log_2(1-p)$$

defined for $p \in (0,1)$. For any fixed p, the entropy function appears in the asymptotic formula

$$\binom{n}{pn} = \frac{n^n e^{-n}}{(pn)^{pn} e^{-pn}((1-p)n)^{(1-p)n} e^{-(1-p)n}} (1+o(1))^n = 2^{n(H(p)+o(1))}.$$

For $p \in (0,.5)$, we further bound

$$\sum_{i \le pn} \binom{n}{i} \le (1+pn)\binom{n}{pn} = 2^{n(H(p)+o(1))}.$$

Theorem 1.1 (Shannon's Theorem). *Let $p \in (0,.5)$ be fixed. For $\epsilon > 0$ arbitrarily small, there exists a coding scheme with rate of transmission greater than $1 - H(p) - \epsilon$ and probability of incorrect transmission less than ϵ.*

Proof. Let $\delta > 0$ be such that $p + \delta < .5$ and $H(p+\delta) < H(p) + \epsilon/2$. For n large, set $m = n(1 - H(p) - \epsilon)$, guaranteeing the rate of transmission. Let $f : \{0,1\}^m \to \{0,1\}^n$ be a random function—each $f(x)$ uniformly and independently chosen. Given f define the decoding function $g : \{0,1\}^n \to \{0,1\}^m$ by setting $g(y) = x$ if x is the unique vector in $\{0,1\}^m$ within $n(p+\delta)$ of $f(x)$. We measure distance by the Hamming metric $\rho : \rho(y,y')$ is the number of coordinates in which y, y' differ. If there is no such x, or more than one such x, then we shall consider decoding to be incorrect.

There are two ways decoding can be incorrect. Possibly, $f(x) + E$ is not within $n(p+\delta)$ of $f(x)$. The distance from $f(x)+E$ to $f(x)$ is simply the number of ones in E which has binomial distribution $B(n,p)$, and so this occurs with probability $o(1)$, in fact, with exponentially small probability. The only other possibility is that there is some $x' \neq x$ with $f(x') \in S$, where S is the set of y' within $n(p+\delta)$ of $f(x)$. Conditioning on the values $f(x), E, f(x')$ is still uniformly distributed over $\{0,1\}^n$ and hence this occurs with probability $|S|2^{-n}$ for any particular x' and thus with total probability at most

$$2^m |S| 2^{-n} < 2^{-n(\epsilon/2 + o(1))} = o(1).$$

The total probability for incorrect decoding from both sources is thus $o(1)$ and, in fact, exponentially small. For n sufficiently large, this is less than ϵ.

The average over all choices of f, x of the probability of incorrect decoding is less than ϵ. Therefore there exists a specific f (hence a specific coding scheme) with probability of incorrect coding less than ϵ. ■

Shannon's theorem, dealing with the intensely practical subject of communications, puts the shortcomings of the probabilistic approach in sharp contrast. Where is the coding scheme? Supposing that a coding scheme may be found, how can encoding and decoding be rapidly processed? A group code is a coding scheme in which the map $f : \{0,1\}^m \to \{0,1\}^n$ is linear, i.e., $f(0) = 0$ and $f(x+x') = f(x) + f(x')$, all calculations modulo 2. (Alternatively, the range of f is a subgroup of $\{0,1\}^n$.) Group codes are of particular interest, in part because of the ease of encoding.

Theorem 1.2. *Let $p \in (0,.5)$ be fixed. For $\epsilon > 0$ arbitrarily small, there exists a group code with rate of transmission greater than $1 - H(p) - \epsilon$ and probability of incorrect transmission less than ϵ.*

Proof. For $1 \le i \le m$, let $u_i \in \{0,1\}^m$ be that vector with a 1 in position i, all other entries 0. Let $f(u_1),\ldots,f(u_m)$ be chosen randomly and independently and then extend f by setting

$$f(\epsilon_1 u_1 + \cdots + \epsilon_m u_m) = \epsilon_1 f(u_1) + \cdots + \epsilon_m u_m.$$

We follow the proof of Shannon's theorem until bounding the probability that $f(x)+E$ lies within $n(p+\delta)$ of $f(x)$. Set $z = x - x' = \epsilon_1 u_1 + \cdots + \epsilon_m u_m$, again all modulo 2. As $x \neq x'$, $z \neq 0$. Reorder for convenience so that $\epsilon_m = 1$. By linearity, $f(z) = f(x) - f(x')$, so we bound $\Pr[f(z) \in S]$, where S is the set of vectors within $n(p+\delta)$ of E. Fixing E and the $f(u_i)$, $i < m$, $f(z)$ still has an additive term $f(u_m)$ that is uniform and independent. Hence $f(z)$ is distributed uniformly. Thus $\Pr[f(z) \in S] = |S|2^{-n}$ and the remainder of the proof is as in Shannon's theorem. ■

2. AN EASY GAME

The (n,k) easychip game is a perfect information game with two players, Pusher and Chooser. There is a board with positions $0,1,\ldots,k$. There are n chips which initially are all at position 0. Each round has two parts. First the Pusher selects two disjoint sets A,B of chips. ($A \cup B$ is not necessarily the set of all chips.) Then Chooser must *either*

- Move every chip of A up one position and eliminate all chips of B.

or

- Move every chip of B up one position and eliminate all chips of A.

"Move up" means that a chip at position i is moved to position $i+1$. Once eliminated, a chip never reappears. Pusher wins if any chip ever reaches position k.

If $n = 2^k$, Pusher may win by, at each move, partitioning the remaining chips into equal sets A,B.

Theorem 2.1. *If $n < 2^k$, then Chooser wins.*

Proof 1. Suppose Chooser announces that he will flip a fair coin each round to determine whether to move A/eliminate B or move B/eliminate A.

Given a strategy for Pusher, let W be the *event* that Pusher wins. For each chip x, let W_x be the event that that chip reaches k so that $W = \vee W_x$. Clearly, $\Pr[W_x] \leq 2^{-k}$, since for x to be moved forward k times, the coin must be flipped correctly k times. Thus

$$\Pr[W] \leq \sum \Pr[W_x] \leq n2^{-k} < 1.$$

The (n,k) easychip game is a finite perfect information game. Either there is a perfect strategy for Pusher or a perfect strategy for Chooser. If there were a perfect strategy for Pusher, then, using it, $\Pr[W] = 1$, a contradiction. Hence there must be a perfect strategy for Chooser. ■

Proof 2. Let a chip at position i have weight $2^{-(k-i)}$ and let the weight of a set A of chips be the sum of their weights. Given A, B, Chooser eliminates the set of higher weight. The surviving chips double their weight, but the total weight of all alive chips does not increase. With $n < 2^k$, the initial total weight is less than 1, so no chip ever reaches k and weight 1. ■

These two proofs are related. The weight of a chip is the probability it will reach k with Chooser flipping coins. (Assume here that the game continues until all chips are either at k or eliminated.) The total weight is then the expected number of chips that will reach k. This is less than 1 initially. In Proof 2, Chooser always selects so that this expectation does not increase. This is related to the method of conditional probabilities discussed in Chapter 15.

3. BALANCING VECTOR GAME

The balancing vector game is a perfect information game with two players, Pusher and Chooser. There is a parameter $n \geq 1$, and we shall be concerned with asymptotics in n. There are n rounds, each involving vectors in R^n. There is a position vector $P \in R^n$, initially set at 0. Each round has two parts. First Pusher picks $v \in \{-1, +1\}^n$. Then Chooser either resets P to $P + v$ or to $P - v$. At the end of the nth round, the payoff to Pusher is $|P|_\infty$, the maximal absolute value of the coordinates of P. Let VAL(n) denote the value of this game to Pusher, and let S_n denote, as usual, the sum of n independent uniform $\{1, -1\}$ random variables.

Theorem 3.1. *If* $\Pr[|S_n| > \alpha] < n^{-1}$, *then* VAL$(n) \leq \alpha$.

Proof. Consider the game a win for Pusher if the final $|P|_\infty > \alpha$. Suppose Chooser announces that he will flip a fair coin each round to determine whether to reset P as $P + v$ or $P - v$. Let x_i be the ith coordinate for the final value of the position vector P. Let W_i be the event $|x_i| > \alpha$ and $W = \vee W_i$, so that W is the event of Pusher winning. Regardless of Pusher's strategy, x_i has distribution S_n, so that

$$\Pr[W] \leq \sum_{i=1}^{n} \Pr[|S_n| > \alpha] < 1.$$

Pusher cannot always win, so Chooser always wins. ■

Corollary 3.2. $\mathrm{VAL}(n) = O(\sqrt{n \ln n})$.

To give a lower bound on $\mathrm{VAL}(n)$, one wants to find a strategy for Pusher that wins against any Chooser. It is not sufficient to find a strategy that does well against a randomly playing Chooser—the Chooser is an adversary. Still, the notion of a randomly playing Chooser motivates the following result.

Theorem 3.3. *If* $\Pr[|S_n| > \alpha] > cn^{-1/2}$, *c an absolute constant, then* $\mathrm{VAL}(n) > \alpha$.

Corollary 3.4. $\mathrm{VAL}(n) = \Omega(\sqrt{n \ln n})$ *and hence* $\mathrm{VAL}(n) = \Theta(\sqrt{n \ln n})$.

Proof of Theorem 3.3. Define, for $x \in Z$, $0 \leq i \leq n$,

$$w_i(x) = \Pr[|x + S_{n-i}| > \alpha].$$

For $P = (x_1, \ldots, x_n)$, set $w_i(P) = \sum_{1 \leq j \leq n} w_i(x_j)$. When P is the position vector at the end of the ith round, $w_i(P)$ may be interpreted as the expected number of coordinates with absolute value greater than α at the end of the game, assuming random play by Chooser. At the beginning of the game, $w_0(P) = w_0(0) > c\sqrt{n}$ by assumption. Given position P at the end of round i, Pusher's strategy will be to select $v \in \{-1, +1\}^n$ so that $w_{i+1}(P - v)$ and $w_{i+1}(P + v)$ are close together.

The distribution $x + S_{n-i}$ splits into $x + 1 + S_{n-i-1}$ and $x - 1 + S_{n-i-1}$ depending on the first coin flip, so that for any i, x,

$$w_i(x) = \tfrac{1}{2}[w_{i+1}(x+1) + w_{i+1}(x-1)].$$

Set $P = (x_1, \ldots, x_n)$, $v = (v_1, \ldots, v_n)$. For $1 \leq j \leq n$, set

$$\Delta_j = w_{i+1}(x_j + 1) - w_{i+1}(x_j - 1)$$

so that

$$w_{i+1}(P + v) - w_{i+1}(P - v) = \sum_{j=1}^{n} v_j \Delta_j,$$

and, for $\epsilon = \pm 1$,

$$w_{i+1}(P + \epsilon v) = w_i(P) + \frac{1}{2}\epsilon \sum_{j=1}^{n} v_j \Delta_j.$$

Now we bound $|\Delta_j|$. Observe that

$$\Delta_j = \Pr[S_{n-i-1} = y] - \Pr[S_{n-i-1} = z],$$

where y is the unique integer of the same parity as $n-i-1$ in the interval $(\alpha-(x_j+1), \alpha-(x_j-1)]$ and z the same in $[-\alpha-(x_j+1), -\alpha-(x_j-1))$. Let us set

$$g(m) = \max_s \Pr[S_m = s] = \binom{m}{\lfloor m/2 \rfloor} 2^{-m} \sim \sqrt{\frac{2}{\pi m}}$$

so that $|\Delta_j| \le g(n-i-1)$ for all j.

A simple strategy for Pusher is then to reorder the coordinates so that $|\Delta_1| \ge \cdots \ge |\Delta_n|$ and then select $v_1, \ldots, v_n \in \{-1, +1\}$ sequentially, giving $v_i \Delta_i$ the opposite sign of $v_1\Delta_1 + \cdots + v_{i-1}\Delta_{i-1}$. (When $i = 1$ or the sum is 0, choose v_i arbitrarily.) This assures

$$|v_1\Delta_1 + \cdots + v_n\Delta_n| \le |\Delta_1| \le g(n-i-1).$$

Let P^i denote the position vector at the end of the ith round and v Pusher's choice for the $i+1$ round. Then regardless of Chooser's choice of $\epsilon = \pm 1$,

$$\begin{aligned} w_{i+1}(P^{i+1}) = w_{i+1}(P^i + \epsilon v) &\ge w_i(P^i) - \frac{1}{2}\left|\sum_{j=1}^{n} v_j \Delta_j\right| \\ &\ge w_i(P^i) - \frac{1}{2} g(n-i-1). \end{aligned}$$

Thus

$$w_n(P^n) \ge w_0(P^0) - \frac{1}{2}\sum_{i=0}^{n-1} g(n-i-1).$$

Simple asymptotics give that the above sum is asymptotic to $(8n/\pi)^{1/2}$. Choosing $c > (2/\pi)^{1/2}$, $w_n(P^n) > 0$. But $w_n(P^n)$ is simply the *number* of coordinates with absolute value greater than α in the final $P = P^n$. This Pusher strategy assures there is more than zero, hence at least one, such coordinate and therefore Pusher wins. ■

4. NONADAPTIVE ALGORITHMS

Let us modify the balancing game of Section 3 by requiring the vectors selected by Pusher to have coordinates 0 and 1 rather than +1 and −1. Let $\mathrm{VAL}^*(n)$ denote the value of the modified game. One can use the bounds on $\mathrm{VAL}(n)$ to show $\mathrm{VAL}^*(n) = \Theta(\sqrt{n \ln n})$.

In Chapter 12, we showed that any family of n sets $S_1, \ldots, S_n$ on n points $1, \ldots, n$ has discrepancy $O(\sqrt{n})$, that is, there is a coloring $\chi : \{1, \ldots, n\} \to \{-1, +1\}$ so that all $|\chi(S_i)| \le c\sqrt{n}$. The proof of this result does not yield an effective algorithm for finding such a coloring and indeed it is not known if there is a polynomial time algorithm to do so. Suppose one asks for a *nonadaptive* algorithm in the following sense. Instead of being presented the entire data of $S_1, \ldots, S_n$ at once, one is presented with the points sequentially. At the jth "round," the algorithm looks at point j—more specifically, at which sets S_i contain j or, equivalently, at the jth column of the incidence matrix. At that stage, the algorithm must decide how to color j and, once colored, the coloring cannot be changed. How small can we assure $\max|\chi(S_i)|$ with such an algorithm? We may think of the points as being presented by an adversary. Thinking of the points as their associated column vectors, Pusher as the worst-case adversary and Chooser as the algorithm, the best such an algorithm can do is precisely $\mathrm{VAL}^*(n)$.

The requirement that an algorithm be nonadaptive is both stronger and weaker than the requirement that an algorithm take polynomial time. Still, this lends support to the conjecture that there is no polynomial time algorithm for finding a coloring with all $|\chi(S_i)| \le c\sqrt{n}$.

The Probabilistic Lens: An Extremal Graph

Let T (top) and B (bottom) be disjoint sets of size m and let G be a bipartite graph, all edges between T and B. Suppose G contains no 4-cycle. How many edges can G have? This is a question from extremal graph theory. Surprisingly, for some m, we may give the precise answer.

Suppose $m = n^2 + n + 1$ and that a projective plane P of order n (and hence containing m points) exists. Identify T with the points of P and B with the lines of P and define $G = G_P$ by letting $t \in T$ be adjacent to $b \in B$ if and only if point t is on line b in P. As two points cannot lie on two lines, G_P contains no 4-cycle. We claim that such a G_P has the largest number of edges of any G containing no 4-cycle and further that any G containing no 4-cycle and having that many edges can be written in the form $G = G_P$.

Suppose G contains no 4-cycle. Let $b_1, b_2 \in B$ be a uniformly selected pair of distinct elements. For $t \in T$, let $D(t)$ be the set of $b \in B$ adjacent to t and $d(t) = |D(t)|$, the degree of t. Let I_t be the indicator random variable for t being adjacent to b_1, b_2. Then

$$E[I_t] = \Pr[b_1, b_2 \in D(t)] = \binom{d(t)}{2} \Big/ \binom{n}{2}.$$

Now set

$$X = \sum_{t \in T} I_t,$$

the number of $t \in T$ adjacent to b_1, b_2. Then $X \leq 1$, that is, all b_1, b_2 have at most one common neighbor. ($X \leq 1$ is actually equivalent to G containing no 4-cycle.) Linearity of expectation gives

$$E[X] = \sum_{t \in T} E[I_t] = \sum_{t \in T} \binom{d(t)}{2} \Big/ \binom{n}{2}.$$

Let $\bar{d} = n^{-1} \sum_{t \in T} d(t)$ be the average degree. Convexity of the function $\binom{y}{2}$ gives

$$\sum_{t \in T} \binom{d(t)}{2} \Big/ \binom{n}{2} \geq n \binom{\bar{d}}{2} \Big/ \binom{n}{2}$$

with equality if and only if all $t \in T$ have the same degree. Now

$$1 \geq \max X \geq E[X] \geq n \binom{\overline{d}}{2} \bigg/ \binom{n}{2}.$$

When $G = G_P$, all $d(x) = \overline{d}$ (every line has $n+1$ points) and $X = 1$ always (two points determine precisely one line), so that the above inequalities are all equalities and

$$1 = n \binom{\overline{d}}{2} \bigg/ \binom{n}{2}.$$

Any graph with more edges would have a strictly larger $\overline{d}$, so that $1 \geq n\binom{\overline{d}}{2}/\binom{n}{2}$ would fail and the graph would contain a 4-cycle.

Suppose further G has the same number of edges as G_P and contains no 4-cycle. The inequalities then must be equalities and so $X = 1$ always. Define a geometry with points T and lines given by the neighbor sets of $b \in B$. As $X = 1$, any two points determine a unique line. Reversing the roles of T, B, one also has that any two lines must determine a unique point. Thus G is generated from a projective plane.

15

Derandomization

As mentioned in Chapter 1, the probabilistic method supplies, in many cases, effective randomized algorithms for various algorithmic problems. In some cases, these algorithms can be derandomized and converted into deterministic ones. In this chapter, we discuss some examples.

1. THE METHOD OF CONDITIONAL PROBABILITIES

An easy application of the basic probabilistic method implies the following statement, which is a special case of Theorem 3.1 in Chapter 2.

Proposition 1.1. *For every integer n, there exists a coloring of the edges of the complete graph K_n by two colors so that the total number of monochromatic copies of K_4 is at most $\binom{n}{4} \cdot 2^{-5}$.*

Indeed, $\binom{n}{4} \cdot 2^{-5}$ is the expected number of monochromatic copies of K_4 in a random 2-edge-coloring of K_n, and hence a coloring as above exists.

Can we actually find *deterministically* such a coloring in time that is polynomial in n? Let us describe a procedure that does it, and is a special case of a general technique called the *method of conditional probabilities*.

We first need to define a weight function for any partially colored K_n. Given a coloring of some of the edges of K_n by red and blue, we define, for each copy K of K_4 in K_n, a weight $w(K)$ as follows. If at least one edge of K is colored red and at least one edge is colored blue, then $W(K) = 0$. If no edge of K is colored, then $w(K) = 2^{-5}$, and if $r \geq 1$ edges of K are colored, all with the same color, then $w(K) = 2^{6-r}$. Also define the total weight W of the partially colored K_n as the sum $\sum w(K)$, as K ranges over all copies of K_4 in K_n. Observe that the weight of each copy K of K_4 is precisely the probability that it will be monochromatic, if all the presently uncolored edges of K_n will be assigned randomly and independently one of the two colors red and blue. Hence, by linearity of expectation, the total weight W is simply the expected number of monochromatic copies of K_4 in such a random extension of the partial coloring of K_n to a full coloring.

We can now describe the procedure for finding a coloring as in Proposition 1.1. Order the $\binom{n}{2}$ edges of K_n arbitrarily, and construct the desired 2-coloring by coloring each edge either red or blue in its turn. Suppose $e_1,\ldots,e_{i-1}$ have already been colored, and we now have to color e_i. Let W be the weight of K_n, as defined above, with respect to the given partial coloring c of $e_1,\ldots,e_{i-1}$. Similarly, let W_{red} be the weight of K_n with respect to the partial coloring obtained from c by coloring e_i red, and let W_{blue} be the weight of K_n with respect to the partial coloring obtained from c by coloring e_i blue. By the definition of W (and as follows from its interpretation as an expected value),

$$W = \frac{W_{\text{red}} + W_{\text{blue}}}{2}.$$

The color of e_i is now chosen so as to minimize the resulting weight, i.e., if $W_{\text{red}} \le W_{\text{blue}}$, then we color e_i red; otherwise, we color it blue. By the above inequality, the weight function never increases during the algorithm. Since at the beginning, its value is exactly $\binom{n}{4}2^{-5}$, its value at the end is at most this quantity. However, at the end all edges are colored, and the weight is precisely the number of monochromatic copies of K_4. Thus the procedure above produces, deterministically and in polynomial time, a 2-edge-coloring of K_n satisfying the conclusion of Proposition 1.1.

Let us describe, now, the method of conditional probabilities in a more general setting. An instance of this method is due, implicitly, to Erdős and Selfridge (1973), and more explicit examples appear in Spencer (1987) and in Raghavan (1988). Suppose we have a probability space, and assume, for simplicity, that it is symmetric and contains 2^l points, denoted by the binary vectors of length l. Let $A_1,\ldots,A_s$ be a collection of events and suppose that $\sum_{i=1}^{s} \Pr(A_i) = k$. Thus k is the expected value of the number of events A_i that hold, and hence there is a point $(\epsilon_1,\ldots,\epsilon_l)$ in the space in which at most k events hold. Our objective is to find such a point deterministically.

For each choice of $(\epsilon_1,\ldots,\epsilon_{j-1})$ and for each event A_i, the conditional probability

$$\Pr(A_i \mid \epsilon_1,\ldots,\epsilon_{j-1})$$

of the event A_i given the values of $\epsilon_1,\ldots,\epsilon_{j-1}$ is obviously the average of the two conditional probabilities corresponding to the two possible choices for ϵ_j. That is,

$$\Pr(A_i \mid \epsilon_1,\ldots,\epsilon_{j-1}) = \frac{\Pr(A_i \mid \epsilon_1,\ldots,\epsilon_{j-1},0) + \Pr(A_i \mid \epsilon_1,\ldots,\epsilon_{j-1},1)}{2}.$$

Consequently,

$$\begin{aligned}
&\sum_{i=1}^{s} \Pr(A_i \mid \epsilon_1,\ldots,\epsilon_{j-1}) \\
&\quad = \frac{\sum_{i=1}^{s} \Pr(A_i \mid \epsilon_1,\ldots,\epsilon_{j-1},0) + \sum_{i=1}^{s} \Pr(A_i \mid \epsilon_1,\ldots,\epsilon_{j-1},1)}{2} \\
&\quad \ge \min\left\{ \sum_{i=1}^{s} \Pr(A_i \mid \epsilon_1,\ldots,\epsilon_{j-1},0), \sum_{i=1}^{s} \Pr(A_i \mid \epsilon_1,\ldots,\epsilon_{j-1},1) \right\}.
\end{aligned}$$

Therefore, if the values of ϵ_j are chosen, each one in its turn, so as to minimize the value of $\sum_{i=1}^{s} \Pr(A_i \mid \epsilon_1,\ldots,\epsilon_j)$, then the value of this sum cannot increase. Since this sum is k at the beginning, it follows that it is at most k at the end. But at the end, each ϵ_j is fixed, and hence the value of this sum is precisely the number of events A_i that hold at the point $(\epsilon_1,\ldots,\epsilon_l)$, showing that our procedure works.

Note that the assumptions that the probability space is symmetric and that it has 2^l points can be relaxed. The procedure above is efficient provided l is not too large (as is usually the case in combinatorial examples), and, more importantly, provided the conditional probabilities $\Pr(A_i \mid \epsilon_1,\ldots,\epsilon_j)$ can be computed efficiently for each of the events A_i and for each possible value of $\epsilon_1,\ldots,\epsilon_j$. This is, indeed, the case in the example considered in Proposition 1.1. However, there are many interesting instances where this is not the case. A trick that can be useful in such cases is the introduction of *pessimistic estimators*, introduced by Raghavan (1987). Consider, again, the symmetric probability space with 2^l points described above, and the events $A_1,\ldots,A_s$ in it. Suppose that for each event A_i, and for each $0 \le j \le l$, we have a function $f_j^i(\epsilon_1,\ldots,\epsilon_j)$, which can be efficiently computed. Assume, also, that

$$f_{j-1}^i(\epsilon_1,\ldots,\epsilon_{j-1}) \ge \min\{f_j^i(\epsilon_1,\ldots,\epsilon_{j-1},0), f_j^i(\epsilon_1,\ldots,\epsilon_{j-1},1)\}, \qquad (1)$$

and that f_j^i is an upper bound on the conditional probabilities for the event A_i, i.e.,

$$f_j^i(\epsilon_1,\ldots,\epsilon_j) \ge \Pr(A_i \mid \epsilon_1,\ldots,\epsilon_j). \qquad (2)$$

In this case, if in the beginning $\sum_{i=1}^{s} f_0^i \le t$, and we choose the values of the ϵ_j so as to minimize the sum $\sum_{i=1}^{s} f_j^i(\epsilon_1,\ldots,\epsilon_j)$ in each step, we get in the end a point $(\epsilon_1,\ldots,\epsilon_l)$ for which the sum $\sum_{i=1}^{s} f_l^i(\epsilon_1,\ldots,\epsilon_l) \le t$. The number of events A_i that hold in this point is at most t. The functions f_j^i in the argument above are called pessimistic estimators.

This enables us to obtain efficient algorithms in some cases where there is no known efficient way of computing the required conditional probabilities. The following theorem is an example; it is related to some of the results in Chapter 14 and Chapter 12.

Theorem 1.2. *Let $(a_{ij})_{i,j=1}^{n}$ be an n by n matrix of reals, where $-1 \le a_{ij} \le 1$ for all i,j. Then one can find, in polynomial time, $\epsilon_1,\ldots,\epsilon_n \in \{-1,1\}$ such that for every i, $1 \le i \le n$, the inequality $|\sum_{j=1}^{n} \epsilon_j a_{ij}| \le \sqrt{2n\ln(2n)}$ holds.*

Proof. Consider the symmetric probability space on the 2^n points corresponding to the 2^n possible vectors $(\epsilon_1,\ldots,\epsilon_n) \in \{-1,1\}^n$. Define $\beta = \sqrt{2n\ln(2n)}$ and let A_i be the event $|\sum_{j=1}^{n} \epsilon_j a_{ij}| > \beta$. We next show that the method of conditional probabilities with appropriate pessimistic estimators enables us to find efficiently a point of the space in which no event A_i holds.

Define $\alpha = \beta/n$ and let $G(x)$ be the function

$$G(x) = \cosh(\alpha x) = \frac{e^{\alpha x} + e^{-\alpha x}}{2}.$$

By comparing the terms of the corresponding Taylor series, it is easy to see that for every real x,

$$G(x) \le \exp(\alpha^2 x^2/2),$$

with strict inequality if both x and α are not 0. It is also simple to check that for every real x and y,

$$G(x)G(y) = \frac{G(x+y) + G(x-y)}{2}.$$

We can now define the functions f_p^i which will form our pessimistic estimators. For each $1 \le i \le n$ and for each $\epsilon_1, \ldots, \epsilon_p \in \{-1, 1\}$, we define

$$f_p^i(\epsilon_1, \ldots, \epsilon_p) = 2e^{-\alpha\beta} G\left(\sum_{j=1}^{p} \epsilon_j a_{ij}\right) \prod_{j=p+1}^{n} G(a_{ij}).$$

Obviously, these functions can be efficiently computed. It remains to check that they satisfy the conditions described in equations (1) and (2), and that the sum $\sum_{i=1}^{n} f_0^i$ is less than 1. This is proved in the following claims. ■

Claim 1.3. *For every $1 \le i \le n$ and every $\epsilon_1, \ldots, \epsilon_{p-1} \in \{-1, 1\}$,*

$$f_{p-1}^i(\epsilon_1, \ldots, \epsilon_{p-1}) \ge \min\{f_p^i(\epsilon_1, \ldots, \epsilon_{p-1}, -1), f_p^i(\epsilon_1, \ldots, \epsilon_{p-1}, 1)\}.$$

Proof. Put $v = \sum_{j=1}^{p-1} \epsilon_j a_{ij}$. By the definition of f_p^i and by the properties of G,

$$\begin{aligned}
f_{p-1}^i(\epsilon_1, \ldots, \epsilon_{p-1}) &= 2e^{-\alpha\beta} G(v) G(a_{ip}) \prod_{j=p+1}^{n} G(a_{ij}) \\
&= 2e^{-\alpha\beta} \frac{G(v - a_{ip}) + G(v + a_{ip})}{2} \prod_{j=p+1}^{n} G(a_{ij}) \\
&= \frac{f_p^i(\epsilon_1, \ldots, \epsilon_{p-1}, -1) + f_p^i(\epsilon_1, \ldots, \epsilon_{p-1}, 1)}{2} \\
&\ge \min\{f_p^i(\epsilon_1, \ldots, \epsilon_{p-1}, -1), f_p^i(\epsilon_1, \ldots, \epsilon_{p-1}, 1)\},
\end{aligned}$$

completing the proof of the claim. ■

Claim 1.4. *For every $1 \le i \le n$ and every $\epsilon_1, \ldots, \epsilon_{p-1} \in \{-1, 1\}$,*

$$f_{p-1}^i(\epsilon_1, \ldots, \epsilon_{p-1}) \ge \Pr(A_i \mid \epsilon_1, \ldots, \epsilon_{p-1}).$$

Proof. Define ν as in the proof of Claim 1.3. Then

$$\begin{aligned}
\Pr(A_i \mid \epsilon_1,\ldots,\epsilon_{p-1}) &\le \Pr\left(\nu + \sum_{j\ge p}\epsilon_j a_{ij} > \beta\right) + \Pr\left(-\nu - \sum_{j\ge p}\epsilon_j a_{ij} > \beta\right)\\
&= \Pr\left(\exp\left(\alpha\left[\nu + \sum_{j\ge p}\epsilon_j a_{ij}\right]\right) > e^{\alpha\beta}\right)\\
&\quad + \Pr\left(\exp\left(-\alpha\left[\nu + \sum_{j\ge p}\epsilon_j a_{ij}\right]\right) > e^{\alpha\beta}\right)\\
&\le e^{\alpha\nu}e^{-\alpha\beta}E\left(\exp\left(\alpha\left[\sum_{j\ge p}\epsilon_j a_{ij}\right]\right)\right)\\
&\quad + e^{-\alpha\nu}e^{-\alpha\beta}E\left(\exp\left(-\alpha\left[\sum_{j\ge p}\epsilon_j a_{ij}\right]\right)\right)\\
&= 2e^{-\alpha\beta}G(\nu)\prod_{j\ge p}G(a_{ij}) = f_{p-1}^i(\epsilon_1,\ldots,\epsilon_{p-1}).
\end{aligned}$$

This completes the proof of Claim 1.4. ■

To establish the theorem, it remains to show that $\sum_{i=1}^n f_0^i < 1$. Indeed, by the properties of G and by the choice of α and β,

$$\begin{aligned}
\sum_{i=1}^n f_0^i &= \sum_{i=1}^n 2e^{-\alpha\beta}\prod_{j=1}^n G(a_{ij})\\
&\le \sum_{i=1}^n 2e^{-\alpha\beta}\prod_{j=1}^n \exp\left(\frac{\alpha^2 a_{ij}^2}{2}\right)\\
&\le \sum_{i=1}^n 2e^{-\alpha\beta}\exp\left(\frac{\alpha^2 n}{2}\right)\\
&= 2n\exp\left(\frac{\alpha^2 n}{2} - \alpha\beta\right) = 2n\exp\left(-\frac{\alpha^2 n}{2}\right) = 1.
\end{aligned}$$

Moreover, the first inequality is strict unless $a_{ij} = 0$ for all i,j, whereas the second is strict unless $a_{ij}^2 = 1$ for all i,j. This completes the proof of the theorem. ■

2. d-WISE INDEPENDENT RANDOM VARIABLES IN SMALL SAMPLE SPACES

The complexity class NC is, roughly speaking, the class of all problems that can be solved in time which is polylogarithmic (in the size of the input) using a polynomial number of parallel processors. Several models of computation, which are a theoretical abstraction of the parallel computer, have been used in considering this class. The most common one is the EREW (= exclusive read, exclusive write) PRAM, in which different processors are not allowed to read from or write into the same memory cell simultanously. See Karp and Ramachandran (1990) for more details.

Let n denote the size of the input. There are several simple tasks that can be easily performed in NC. For example, it is possible to copy the content of a cell c into $m = n^{O(1)}$ cells in time $O(\log n)$, using, say, m processors. To do so, consider a complete binary tree with m leaves and associate each of its internal vertices with a processor. At first, the processor corresponding to the root of the tree reads from c and writes its content in two cells, corresponding to its two children. Next, each of these two, in parallel, reads from its cell and writes its content in two cells corresponding to its two children. In general, at the ith step all the processors whose distance from the root of the tree is $i - 1$, in parallel, read the content of c previously stored in their cells and write it twice. The procedure clearly ends in time $O(\log m)$, as claimed. (In fact, it can be shown that $O(m/\log m)$ processors suffice for this task, but we do not try to optimize this number here.)

A similar technique can be used for computing the sum of m numbers with m processors in time $O(\log m)$: we consider the numbers as if they lie on the leaves of a complete binary tree with m leaves, and in the ith step each one of the processors whose distance from the leaves is i computes, in parallel, the sum of the two numbers previously computed by its children. The root will clearly have, in such a way, the desired sum in time $O(\log m)$.

Let us now return to the edge-coloring problem of the complete graph K_n discussed in Proposition 1.1. By the remarks above, the problem of *checking*, if in a given edge-coloring there are at most $\binom{n}{4}2^{-5}$ monochromatic copies of K_4, is in NC, i.e., this checking can be done in time $(\log n)^{O(1)}$—in fact, in time $O(\log n)$—using $n^{O(1)}$ processors. Indeed, we can first copy the given coloring $\binom{n}{4}$ times. Then we assign a processor for each copy of K_4 in K_n, and this processor checks if its copy is monochromatic or not (all these checkings can be done in parallel, since we have enough copies of the coloring). Finally, we sum the number of processors whose copies are monochromatic. Clearly, we can complete the work in time $O(\log n)$ using $n^{O(1)}$ parallel processors.

Thus we can *check*, in NC, if a given coloring of K_n satisfies the assertion of Proposition 1.1. Can we *find* such a coloring deterministically in NC? The method described in the previous section does not suffice, as the edges have been colored one by one, so the procedure is sequential and requires time $\Omega(n^2)$. However, it turns out that in fact we can find, in NC, a coloring with the desired properties by applying a method that relies on a technique first suggested by Joffe (1974), and later developed by many researchers. This method is a general technique for converting randomized algorithms whose analysis only depends on d-wise rather than fully independent random choices (for some constant d) into deterministic (and in many cases also parallel) ones. Our approach here follows the one of Alon, Babai, and Itai (1986), but for simplicity we only consider here the case of random variables that take the two values $0, 1$ with equal probability.

The basic idea is to replace an exponentially large sample space by one of polynomial size. If a random variable on such a space takes a certain value with positive probability, then we can find a point in the sample space in which this happens simply by deterministically checking all the points. This can be done with no loss of time by using a polynomial number of parallel processors. Note that for the edge-coloring problem considered in Proposition 1.1, 6-wise independence of the random variables corresponding to the colors of the edges suffices—since this already gives a probability of 2^{-5} for each copy of K_4 to be monochromatic, and hence gives the required expected value of monochromatic copies. Therefore, for this specific example, it suffices to construct a sample space of size $n^{O(1)}$ and $\binom{n}{2}$ random variables in it, each taking the values 0 and 1 with probability 1/2, such that each 6 of the random variables are independent.

Small sample spaces with many d-wise independent $0,1$-random variables in them can be constructed from any linear error correcting code with appropriate parameters. The construction we describe here is based on the binary BCH codes (see, e.g., MacWilliams and Sloane [1977]).

Theorem 2.1. *Suppose $n = 2^k - 1$ and $d = 2t + 1$. Then there exists a symmetric probability space Ω of size $2(n+1)^t$ and d-wise independent random variables $y_1, \ldots, y_n$ over Ω, each of which takes the values 0 and 1 with probability 1/2.*

The space and the variables are explicitly constructed, given a representation of the field $F = GF(2^k)$ as a k-dimensional algebra over $GF(2)$.

Proof. Let $x_1, \ldots, x_n$ be the n nonzero elements of F, represented as column vectors of length k over $GF(2)$. Let H be the following $1 + kt$ by n matrix over $GF(2)$:

$$\begin{bmatrix} 1 & 1 & \cdots & 1 \\ x_1 & x_2 & \cdots & x_n \\ x_1^3 & x_2^3 & \cdots & x_n^3 \\ & \vdots & & \vdots \\ x_1^{2t-1} & x_2^{2t-1} & \cdots & x_n^{2t-1} \end{bmatrix}$$

This is the parity check matrix of the extended binary BCH code of length n and designed distance $2t + 2$. It is well known that any $d = 2t + 1$ columns of H are linearly independent over $GF(2)$. For completeness, we present the proof in the next lemma. ■

Lemma 2.2. *Any set of $d = 2t + 1$ columns of H is linearly independent over $GF(2)$.*

Proof. Let $J \subset \{1,2,\ldots,n\}$ be a subset of cardinality $|J| = 2t + 1$ of the set of indices of the columns of H. Suppose that $\sum_{j \in J} z_j H_j = 0$, where H_j denotes the jth column of H and $z_j \in GF(2)$. To complete the proof, we must show that $z_j = 0$ for all $j \in J$. By the assumption,

$$\sum_{j \in J} z_j x_j^i = 0 \tag{3}$$

for $i = 0$ and for every odd i satisfying $1 \leq i \leq 2t - 1$. Suppose, now, that $a = 2^b \cdot l$, where $l \leq 2t - 1$ is an odd number. By squaring equation (3) for $i = l$ b times, using the fact that in characteristic 2, $(u + v)^2 = u^2 + v^2$, and the fact that since each z_j is either 0 or 1, the equality $z_j = z_j^2$ holds for all j, we conclude that (3) holds for $i = a$. Consequently, (3) holds for all i, $0 \leq i \leq 2t$. This is a homogeneous system of $2t + 1$ linear equations in $2t + 1$ variables. The matrix of the coefficients is a Vandermonde matrix, which is nonsingular. Thus, the only solution is the trivial one $z_j = 0$ for all $j \in J$, completing the proof of the lemma. ■

Proof of Theorem 2.1 (Continued). Returning to the proof of the theorem, we define $\Omega = \{1,2,\ldots,2(n+1)^t\}$, and let $A = (a_{ij})$, $i \in \Omega$, $1 \leq j \leq n$, be the $(0,1)$-matrix whose $2(n+1)^t = 2^{kt+1}$ rows are all the linear combinations (over $GF(2)$) of the rows of H. The sample space Ω is now endowed with the uniform probability measure, and the random variable y_j is defined by the formula $y_j(i) = a_{ij}$ for all $i \in \Omega, 1 \leq j \leq n$.

It remains to show that the variables y_j are d-wise independent, and that each of them takes the values 0 and 1 with equal probability. For this, we have to show that for every set J of up to d columns of A, the rows of the $|\Omega|$ by $|J|$ submatrix $A_J = (a_{ij})$, $i \in \Omega$, $j \in J$ take on each of the $2^{|J|}$ $(0,1)$-vectors of length $|J|$ equally often. However, by Lemma 2.2, the columns of the corresponding submatrix H_J of H are linearly independent. The number of rows of A_J that are equal to any given vector is precisely the number of linear combinations of the rows of H_J that are equal to this vector. This number is the number of solutions of a system of $|J|$ linearly independent linear equations in $kt + 1$ variables, which is, of course, $2^{kt+1-|J|}$, independent of the vector of free coefficients. This completes the proof of the theorem. ■

Theorem 2.1 supplies an efficient way of constructing, for every fixed d and every n, a sample space of size $O(n^{\lfloor d/2 \rfloor})$ and n d-wise independent random variables in it, each taking the values 0 and 1 with equal probability. In particular, we can use such a space of size $O(\binom{n}{2}^3) = O(n^6)$ for finding a coloring as in Proposition 1.1 in NC. Several other applications of Theorem 2.1 appear in the paper of Alon, Babai, and Itai (1986).

It is natural to ask if the size $O(n^{\lfloor d/2 \rfloor})$ can be improved. We next show that this size is optimal, up to a constant factor (depending on d).

Let us call a random variable *almost constant* if it attains a single value with probability 1. Let $m(n,d)$ denote the function defined by

$$m(n,d) = \sum_{j=0}^{d/2} \binom{n}{j} \qquad \text{if } d \text{ is even,}$$

and

$$m(n,d) = \sum_{j=0}^{(d-1)/2} \binom{n}{j} + \binom{n-1}{(d-1)/2} \qquad \text{if } d \text{ is odd.}$$

Observe that for every fixed d, $m(n,d) = \Omega(n^{\lfloor d/2 \rfloor})$.

Proposition 2.3. *If the random variables $y_1, \ldots, y_n$ over the sample space Ω are d-wise independent and none of them is almost constant, then $|\Omega| \geq m(n,d)$.*

Note that we assume here neither that Ω is a symmetric space nor that the variables y_j are $(0,1)$-variables.

Proof. Clearly, we may assume that the expected value of each y_j is 0 (since otherwise we can replace y_j by $y_j - E(y_j)$). For each subset S of $\{1, \ldots, n\}$, define $\alpha_S = \prod_{j \in S} y_j$. Observe that since no y_j is almost constant and since the variables are d-wise independent,

$$E(\alpha_S \alpha_S) = \prod_{j \in S} \mathrm{var}(y_j) > 0 \tag{4}$$

for all S satisfying $|S| \leq d$. Similarly, for all S and T satisfying $|S \cup T| \leq d$ and $S \neq T$, we have

$$E(\alpha_s \alpha_T) = \prod_{j \in S \cap T} \mathrm{var}(y_j) \prod_{j \in S \cup T \setminus (S \cap T)} E(y_j) = 0. \tag{5}$$

Let $S_1, \ldots S_m$, where $m = m(n,d)$, be subsets of $\{1, \ldots, n\}$ such that the union of each two is of size at most d. (Take all subsets of size at most $d/2$, and if d is odd add all the subsets of size $(d+1)/2$ containing 1.)

To complete the proof, we show that the m functions α_{S_j} (considered as real vectors of length $|\Omega|$) are linearly independent. This implies that $|\Omega| \geq m = m(n,d)$, as stated in the proposition.

To prove linear independence, suppose $\sum_{j=1}^{m} c_j \alpha_{S_j} = 0$. Multiplying by α_{S_i} and computing expected values we obtain, by (5),

$$0 = \sum_{j=1}^{m} c_j E(\alpha_{S_j} \alpha_{S_i}) = c_i E(\alpha_{S_i} \alpha_{S_i}).$$

This implies, by (4), that $c_i = 0$ for all i. The required linear independence follows, completing the proof. ■

The last proposition shows that the size of a sample space with n d-wise independent nontrivial random variables can be polynomial in n only when d is fixed. However, as shown by Naor and Naor (1990), if we only require the random variables to be *almost* d-wise independent, the size can be polynomial even when $d = \Omega(\log n)$. Such sample spaces and random variables, that can be constructed explicitly in several ways, have various interesting applications in which almost d-wise independence suffices. More details appear in Naor and Naor (1990) and in Alon, et al. (1990).

APPENDIX A

Bounding of Large Deviations

We give here some basic bounds on large deviations that are useful when employing the probabilistic method. Our treatment is self-contained. Most of the results may be found in, or immediately derived from, the seminal paper of H. Chernoff (1952). While we are guided by asymptotic considerations, the inequalities are proven for all values of the parameters in the specified region. The first result, while specialized, contains basic ideas found throughout the appendix.

Theorem A.1. *Let $X_i, 1 \leq i \leq n$, be mutually independent random variables with*

$$\Pr[X_i = +1] = \Pr[X_i = -1] = \tfrac{1}{2}$$

and set, following the usual convention,

$$S_n = X_1 + \cdots X_n.$$

Let $a > 0$. Then

$$\Pr[S_n > a] < e^{-a^2/2n}.$$

Remark. For large n, the central limit theorem implies that S_n is approximately normal with zero mean and standard deviation $\sqrt{n}$. In particular, for any fixed u,

$$\lim_{n\to\infty} \Pr\left[S_n > u\sqrt{n}\right] = \int_{t=u}^{\infty} \frac{1}{\sqrt{2\pi}} e^{-t^2/2}\, dt,$$

which one can show directly is less than $e^{-u^2/2}$. Our proof, we emphasize once again, is valid for all n and all $a > 0$.

We require Markov's inequality, which states: Suppose that Y is an arbitrary nonnegative random variable, $\alpha > 0$. Then

$$\Pr[Y > \alpha E[Y]] < \frac{1}{\alpha}.$$

Proof of Theorem A.1. Fix n, a, and let, for the moment, $\lambda > 0$ be arbitrary. For $1 \leq i \leq n$,

$$E[e^{\lambda X_i}] = \frac{e^{\lambda} + e^{-\lambda}}{2} = \cosh(\lambda).$$

We require the inequality

$$\cosh(\lambda) \leq e^{\lambda^2/2},$$

valid for all $\lambda > 0$, the special case $\alpha = 0$ of Lemma A.5 below. (The inequality may be more easily shown by comparing the Taylor series of the two functions termwise.)

$$e^{\lambda S_n} = \prod_{i=1}^{n} e^{\lambda X_i}.$$

Since the X_i are mutually independent, so are the $e^{\lambda X_i}$, expectations multiply and

$$E[e^{\lambda S_n}] = \prod_{i=1}^{n} E[e^{\lambda X_i}] = [\cosh(\lambda)]^n < e^{\lambda^2 n/2}.$$

We note that $S_n > a$ if and only if $e^{\lambda S_n} > e^{\lambda a}$ and apply Markov's inequality so that

$$\Pr[S_n > a] = \Pr[e^{\lambda S_n} > e^{\lambda a}] \leq E[e^{\lambda S_n}]/e^{\lambda a} \leq e^{\lambda^2 n/2 - \lambda a}.$$

We set $\lambda = a/n$ to optimize the inequality: $\Pr[S_n > a] < e^{-a^2/2n}$, as claimed. ■

By symmetry, we immediately have

Corollary A.2. *Under the assumptions of Theorem* A.1,

$$\Pr[|S_n| > a] < 2e^{-a^2/2n}.$$

Our remaining results will deal with distributions X of the following prescribed type.

Assumptions A.3.

$$p_1, \ldots, p_n \in [0,1]$$

$$p = \frac{p_1 + \ldots + p_n}{n}$$

$X_1, \ldots, X_n$ mutually independent with

$$\Pr[X_i = 1 - p_i] = p_i$$

$$\Pr[X_i = -p_i] = 1 - p_i$$

$$X = X_1 + \cdots + X_n.$$

Remark. Clearly, $E[X] = E[X_i] = 0$. When all $p_i = 1/2$, X has distribution $S_n/2$. When all $p_i = p$, X has distribution $B(n,p) - np$, where $B(n,p)$ is the usual binomial distribution.

Theorem A.4. *Under Assumptions* A.3 *and with* $a > 0$,

$$\Pr[X > a] < e^{-2a^2/n}.$$

Lemma A.5. *For all reals* α, β *with* $|\alpha| \le 1$,

$$\cosh(\beta) + \alpha \sinh(\beta) \le e^{\beta^2/2 + \alpha\beta}.$$

Proof. This is immediate if $\alpha = +1$ or $\alpha = -1$ or $|\beta| \ge 100$. If the lemma were false, the function

$$f(\alpha,\beta) = \cosh(\beta) + \alpha\sinh(\beta) - e^{\beta^2/2 + \alpha\beta}$$

would assume a negative global minimum in the interior of the rectangle

$$R = \{(\alpha,\beta) : |\alpha| \le 1,\ |\beta| \le 100\}.$$

Setting partial derivatives equal to 0, we find

$$\sinh(\beta) + \alpha\cosh(\beta) = (\alpha + \beta)e^{\beta^2/2 + \alpha\beta},$$

$$\sinh(\beta) = \beta e^{\beta^2/2 + \alpha\beta},$$

and thus $\tanh(\beta) = \beta$, which implies $\beta = 0$. But $f(\alpha, 0) = 0$ for all α, a contradiction. ■

Lemma A.6. *For all* $\theta \in [0,1]$ *and all* λ,

$$\theta e^{\lambda(1-\theta)} + (1-\theta)e^{-\lambda\theta} \le e^{\lambda^2/8}.$$

Proof. Setting $\theta = (1+\alpha)/2$ and $\lambda = 2\beta$, Lemma A.6 reduces to Lemma A.5. ■

Proof of Theorem A.4. Let, for the moment, $\lambda > 0$ be arbitrary.

$$E[e^{\lambda X_i}] = p_i e^{\lambda(1-p_i)} + (1-p_i)e^{-\lambda p_i} \le e^{\lambda^2/8}$$

by Lemma A.6. Then

$$E[e^{\lambda X}] = \prod_{i=1}^{n} E[e^{\lambda X_i}] \le e^{\lambda^2 n/8}.$$

Applying Markov's inequality,

$$\Pr[X > a] = \Pr[e^{\lambda X} > e^{\lambda a}] \leq \frac{E[e^{\lambda X}]}{e^{\lambda a}} \leq e^{\lambda^2 n/8 - \lambda a}.$$

We set $\lambda = 4a/n$ to optimize the inequality: $\Pr[X > a] < e^{-2a^2/n}$, as claimed. ■

Again by symmetry, we immediately have

Corollary A.7. *Under Assumptions* A.3 *and with* $a > 0$,

$$\Pr[|X| > a] < 2e^{-2a^2/n}.$$

Under Assumptions A.3 with $\lambda > 0$ arbitrary,

$$E[e^{\lambda X}] = \prod_{i=1}^{n} E[e^{\lambda X_i}] = \prod_{i=1}^{n} [p_i e^{\lambda(1-p_i)} + (1-p_i)e^{-\lambda p_i}]$$

$$= e^{-\lambda p n} \prod_{i=1}^{n} [p_i e^{\lambda} + (1-p_i)].$$

With λ fixed, the function

$$f(x) = \ln[xe^{\lambda} + 1 - x] = \ln[Bx + 1] \qquad \text{with} \quad B = e^{\lambda} - 1$$

is concave and hence (Jansen's inequality)

$$\sum_{i=1}^{n} f(p_i) \leq nf(p).$$

Exponentiating both sides,

$$\prod_{i=1}^{n} [p_i e^{\lambda} + (1-p_i)] \leq [pe^{\lambda} + (1-p)]^n,$$

so that

Lemma A.8. *Under Assumptions* A.3,

$$E[e^{\lambda X}] \leq e^{-\lambda p n}[pe^{\lambda} + (1-p)]^n.$$

Theorem A.9. *Under Assumptions* A.3 *and with* $a > 0$,

$$\Pr[X > a] < e^{-\lambda p n}[pe^{\lambda} + (1-p)]^n e^{-\lambda a}$$

for all $\lambda > 0$.

Proof. $\Pr[X > a] = \Pr[e^{\lambda X} > e^{\lambda a}] \leq E[e^{\lambda X}]/e^{\lambda a}$. Now apply Lemma A.8. ■

Remark. For given p, n, a, an optimal assignment of λ in Theorem A.9 is found by elementary calculus to be

$$\lambda = \ln\left[\left(\frac{1-p}{p}\right)\left(\frac{a+np}{n-(a+np)}\right)\right].$$

This value is oftentimes too cumbersome to be useful. We employ suboptimal λ to achieve more convenient results.

Setting $\lambda = \ln[1 + a/pn]$ and using the fact that $(1 + a/n)^n \leq e^a$, Theorem A.9 implies

Corollary A.10.

$$\Pr[X > a] < e^{a - pn\ln(1+a/pn) - a\ln(1+a/pn)}.$$

Theorem A.11.

$$\Pr[X > a] < e^{-a^2/2pn + a^3/2(pn)^2}.$$

Proof of Theorem A.11. With $u = a/pn$, apply the inequality

$$\ln(1+u) \geq u - \frac{u^2}{2},$$

valid for all $u \geq 0$, to Corollary A.10. ■

When all $p_i = p$, X has variance $np(1-p)$. With $p = o(1)$ and $a = o(pn)$, this bound reflects the approximation of X by a normal distribution with variance $\sim np$. The bound of Theorem A.11 hits a minimum at $a = 2pn/3$. For $a > 2pn/3$, we have the simple bound

$$\Pr[X > a] \leq \Pr[X > 2pn/3] < e^{-2pn/27}.$$

This is improved by the following.

Theorem A.12. *For* $\beta \geq 1$,

$$\Pr[X \geq (\beta - 1)pn] < [e^{\beta-1}\beta^{-\beta}]^{pn}.$$

Proof. Direct "plug in" to Corollary A.10. ■

$X + pn$ may be interpreted as the number of successes in n independent trials when the probability of success in the ith trial is p_i.

Theorem A.13. *Under Assumptions* A.3 *and with* $a > 0$,

$$\Pr[X < -a] < e^{-a^2/2pn}.$$

Note that one cannot simply employ "symmetry," as then the roles of p and $1-p$ are interchanged.

Proof. Let $\lambda > 0$ be, for the moment, arbitrary. Then

$$E[e^{-\lambda X}] = \prod_{i=1}^{n} E[e^{-\lambda X_i}] = \prod_{i=1}^{n}[p_i e^{-\lambda(1-p_i)} + (1-p_i)e^{\lambda p_i}]$$

$$= e^{\lambda pn}\prod_{i=1}^{n}[p_i e^{-\lambda} + (1-p_i)].$$

With λ fixed, the function

$$f(x) = \ln[xe^{-\lambda} + (1-x)] = \ln[Bx+1] \qquad \text{with} \quad B = e^{-\lambda} - 1$$

is concave. (That B is here negative is immaterial.) Thus

$$\sum_{i=1}^{n} f(p_i) \le nf(p).$$

Exponentiating both sides gives

$$E[e^{-\lambda X}] \le e^{\lambda pn}[pe^{-\lambda} + (1-p)]^n,$$

analogous to Theorem A.8. Then

$$\Pr[X < -a] = \Pr[e^{-\lambda X} > e^{\lambda a}] < e^{\lambda pn}[pe^{-\lambda} + (1-p)]^n e^{-\lambda a},$$

analogous to Theorem A.9. We employ the inequality

$$1 + u \le e^u,$$

valid for all u, so that

$$pe^{-\lambda} + (1-p) = 1 + (e^{-\lambda} - 1)p < e^{p(e^{-\lambda}-1)}$$

and

$$\Pr[X < -a] \le e^{\lambda pn + np(e^{-\lambda}-1) - \lambda a} = e^{np(e^{-\lambda}-1+\lambda) - \lambda a}.$$

We employ the inequality

$$e^{-\lambda} \le 1 - \lambda + \frac{\lambda^2}{2},$$

valid for all $\lambda > 0$. (Note: The analogous inequality $e^{\lambda} \leq 1 + \lambda + \lambda^2/2$ is *not* valid for $\lambda > 0$ and so this method, when applied to $\Pr[X > a]$, requires an "error" term as the one found in Theorem A.11.) Now

$$\Pr[X < -a] \leq e^{np\lambda^2/2 - \lambda a}.$$

We set $\lambda = a/np$ to optimize the inequality:

$$\Pr[X < -a] < e^{-a^2/2pn},$$

as claimed. ■

For clarity the following result is often useful.

Corollary A.14. *Let Y be the sum of mutually independent indicator random variables, $\mu = E[Y]$. For all $\epsilon > 0$,*

$$\Pr[|Y - \mu| > \epsilon\mu] < 2e^{-c_\epsilon \mu},$$

where $c_\epsilon > 0$ depends only on ϵ.

Proof. Apply Theorems A.12 and A.13 with $Y = X + pn$ and

$$c_\epsilon = \min\left[-\ln(e^{\epsilon}(1+\epsilon)^{-(1+\epsilon)}), \frac{\epsilon^2}{2}\right].$$ ■

The asymmetry between $\Pr[X < a]$ and $\Pr[X > a]$ given by Theorems A.12 and A.13 is real. The estimation of X by a normal distribution with zero mean and variance np is roughly valid for estimating $\Pr[X < a]$ for any a and for estimating $\Pr[X > a]$ while $a = o(np)$. But when a and np are comparable or when $a \gg np$, the Poisson behavior "takes over" and $\Pr[X > a]$ *cannot* be accurately estimated by using the normal distribution.

We conclude with two large deviation results involving distributions other than sums of indicator random variables.

Theorem A.15. *Let P have Poisson distribution with mean μ. For $\epsilon > 0$,*

$$\Pr[P \leq \mu(1-\epsilon)] \leq e^{-\epsilon^2\mu/2},$$

$$\Pr[P \geq \mu(1+\epsilon)] \leq \left[e^{\epsilon}(1+\epsilon)^{-(1+\epsilon)}\right]^{\mu}.$$

Proof. For any s,

$$\Pr[P = s] = \lim_{n\to\infty} \Pr\left[B\left(n, \frac{\mu}{n}\right) = s\right].$$

Apply Theorems A.12 and A.13. ■

Theorem A.16. *Let X_i, $1 \le i \le n$, be mutually indpendent with all $E[X_i] = 0$ and all $|X_i| \le 1$. Set $S = X_1 + \cdots + X_n$. Then*

$$\Pr[S > a] < e^{-a^2/2n}.$$

Proof. Set, as in the proof of Theorem A.1, $\lambda = a/n$. Set

$$h(x) = \frac{e^{\lambda} + e^{-\lambda}}{2} + \frac{e^{\lambda} - e^{-\lambda}}{2} x.$$

For $x \in [-1, 1]$, $e^{\lambda x} \le h(x)$. ($y = h(x)$ is the chord through the points $x = \pm 1$ of the convex curve $y = e^{\lambda x}$.) Thus

$$E[e^{\lambda X_i}] \le E[h(X_i)] = h(E[X_i]) = h(0) = \cosh \lambda.$$

The remainder of the proof follows as in Theorem A.1. ■

APPENDIX B

Some Open Problems

By Paul Erdős

Mathematical Institute of the Hungarian Academy of Sciences
Budapest, Hungary

Denote by $r(k,n)$ the smallest integer for which every graph of $r(k,n)$ vertices either contains an independent set of k vertices or a complete graph of n vertices. Let us first assume $k = n$.

$$c_1 n 2^{n/2} < r(n,n) < \frac{\binom{2n}{n}}{n^\alpha} \tag{1}$$

are the best current results, where $\alpha > 0$ is an absolute constant. $r(5,5)$ is not known at present; the best known results are $43 \le r(5,5) \le 55$. The lower bound in (1) is an easy application of the probability method; only the value of c_1 has been improved. The first problem would be to prove

$$\frac{r(n,n)}{n2^{n/2}} \to \infty. \tag{2}$$

Perhaps (2) will not be very difficult. The real problem is to compute

$$\lim_{n\to\infty} r(n,n)^{1/n} = c. \tag{3}$$

It is not even known if the limit in (3) exists. It is not at all certain that random methods will help here.

$$c_2 \frac{n^2}{(\log n)^2} < r(3,n) < c_3 \frac{n^2}{\log n}. \tag{4}$$

Both the upper and the lower bound in (4) are proved by random methods. It would be very nice to improve (4) and perhaps get an asymptotic formula for $r(3,n)$. This might be very difficult and perhaps probability methods will be of no help. I am fairly sure that for any fixed k,

$$r(k,n) > n^{k-1-\epsilon} \tag{5}$$

APPENDIX B

Some Open Problems

By Paul Erdős
Mathematical Institute of the Hungarian Academy of Sciences
Budapest, Hungary

Denote by $r(k,n)$ the smallest integer for which every graph of $r(k,n)$ vertices either contains an independent set of k vertices or a complete graph of n vertices. Let us first assume $k = n$.

$$c_1 n 2^{n/2} < r(n,n) < \frac{\binom{2n}{n}}{n^{\alpha}} \tag{1}$$

are the best current results, where $\alpha > 0$ is an absolute constant. $r(5,5)$ is not known at present; the best known results are $43 \leq r(5,5) \leq 55$. The lower bound in (1) is an easy application of the probability method; only the value of c_1 has been improved. The first problem would be to prove

$$\frac{r(n,n)}{n 2^{n/2}} \rightarrow \infty. \tag{2}$$

Perhaps (2) will not be very difficult. The real problem is to compute

$$\lim_{n\to\infty} r(n,n)^{1/n} = c. \tag{3}$$

It is not even known if the limit in (3) exists. It is not at all certain that random methods will help here.

$$c_2 \frac{n^2}{(\log n)^2} < r(3,n) < c_3 \frac{n^2}{\log n}. \tag{4}$$

Both the upper and the lower bound in (4) are proved by random methods. It would be very nice to improve (4) and perhaps get an asymptotic formula for $r(3,n)$. This might be very difficult and perhaps probability methods will be of no help. I am fairly sure that for any fixed k,

$$r(k,n) > n^{k-1-\epsilon} \tag{5}$$

and believed that the proof for $k > 3$ would not be very difficult and that the difficulties are only technical. It is quite possible that I was wrong and

$$r(4,n) > n^{3-\epsilon}$$

is very hard and may require new methods and perhaps is not even true. The best known result is due to Spencer:

$$r(4,n) > n^{5/2-\epsilon},$$

proved by the Lovász local lemma. One would expect

$$r(k,n) > \frac{n^{k-1}}{(\log n)^{c_k}}.$$

The probabilistic methods of Ajtai–Komlós–Szemerédi give

$$r(k,n) < c\frac{n^{k-1}}{(\log n)^{\alpha_k}}.$$

An asymptotic formula for $r(k,n)$ seems out of reach at present.

Two related problems are: Let $G(n)$ be a triangle-free graph of n vertices. How large can its chromatic number be? $n^{1/2}/(\log n)^{\alpha}$ is a good upper and lower bound, but the best value of α is not known. Let G_e be a triangle-free graph of e edges. How large can its chromatic number be? $e^{1/3}/(\log e)^{c}$ is roughly the correct order of magnitude.

One of the early triumphs of the probability method was to prove that there is a graph of arbitrarily large chromatic number and arbitrarily large girth; more precisely, for every k, there is a graph of n vertices, girth k, the largest independent set of which is $n^{1-\epsilon_k}$ where $\epsilon_k > c/k$. This holds for every k. Lovász gave a complicated construction for a graph of arbitrary large girth and chromatic number; later Nesetril and Rödl gave a much simpler construction. As observed by Alon, some of the known explicit constructions of expanders supply, for every k, explicit graphs of n vertices, with girth k, and with largest independent set $n^{1-\epsilon_k}$ where $\epsilon_k > c/k$. This essentially matches the result obtained by the probabilistic method.

I often asked the following question. Let G be a random graph of n vertices and cn edges. What can be said about the chromatic number of it? What is the largest $r = f(c)$ for which the probability that the chromatic number is r is positive? As far as I know, this is open except for $r = 3$. Also, the graph of chromatic number $r > 4$ will have size $c_r n$. Bollobás had the following idea. Let l be the largest integer for which our graph contains an induced subgraph each vertex of which has degree at least l. Is it true that the chromatic number of this graph is $l+1$? Or, if not, it should be at least l. I think this problem is still open. Its relevance to my problem is clear.

When Rényi and I developed our theory of random graphs, we thought of extending our study for hypergraphs. We mistakenly thought that all (or most) of the extensions would be routine and we completely overlooked the following beautiful question of Shamir. Rényi and I proved that if n is even and we consider the random graph of $\frac{1}{2}n\log n + cn$ edges, then G has a perfect matching with probability $f(c)$, where $f(c)$ is an exponential function in c. We used the theorem of Tutte. Now Shamir asked how many triples must one choose on $3n$ elements so that with probability bounded away from zero one should get n vertex disjoint triples. Shamir proved that $n^{3/2}$ triples suffice, but the truth may very well be $n^{1+\epsilon}$ or even $cn\log n$. The reasons for the difficulty is that Tutte's theorems seem to have no analogy for triple systems or more generally for hypergraphs. Clearly, many related questions can be asked for hypergraphs for all r-tuples, $r \geq 3$. But perhaps for random graphs, there might be also unexplored questions, e.g., how many edges are needed in a graph of $3n$ vertices to be able to cover the vertices by n vertex disjoint triangles? The correct answer will be probably about $n^{4/3}$ edges, but perhaps a little more will be needed. Also instead of a triangle one could ask for other graphs, e.g. the graph of six vertices and six edges with a triangle and a 1 factor. Again, the lack of analogs to Tutte's theorem may cause serious trouble.

Has the following question of Spencer and myself been really settled? We have n vertices and draw edges at random one by one. We stop as soon as there is no isolated vertex. Is it true that then there already almost surely is a perfect matching if the number of vertices is even? If we stop when every vertex has degree 2, is it true that we already have a Hamiltonian circuit? Similarly, if every vertex has degree r, does the graph contain an r-chromatic (or $r+1$ chromatic) subgraph? Here, of course, surely an r chromatic subgraph might appear once there is a subgraph every vertex of which has degree $\geq r-1$. Also, if the first r-chromatic subgraph appears, its size will be very large almost surely if $r > 2$. For $r = 2$ this is not true; the expected value is very large because of the large odd circuits. This has been computed by a Scandinavian mathematician and also by Knuth in a paper, if I remember right, dedicated to my 75th birthday.

How accurately can one estimate the chromatic number of the random graph (with edge probability 1/2)? Perhaps one can prove that the error is more (much more) than $O(1)$. (Shamir and Spencer have an $O(n^{1/2})$ upper bound.) Also, in the proof of Bollobás, is it true that every subgraph of size n^{α} will almost surely contain an independent set of size $c\log n$ with the right value of c?

As stated before, if a random graph of an even number of vertices has enough edges so that every vertex has degree ≥ 1, then there is with probability tending to 1 a perfect matching. Could this hold for other configurations too? That is, if every vertex is contained in a triangle, is there a set of vertex disjoint triangles covering every vertex? Many related questions can be asked.

References

M. Ajtai (1983), Σ_1^1-formulae on finite structures, *Annals of Pure and Applied Logic* **24**, 1–48.

M. Ajtai, J. Komlós, and E. Szemerédi (1983), Sorting in $c \log n$ parallel steps, *Combinatorica* **3**, 1–19.

——— (1987), Deterministic simulation in LOGSPACE, *Proceedings of the 19th Annual ACM STOC*, ACM Press, New York, 132–140.

R. Ahlswede and D.E. Daykin (1978), An inequality for the weights of two families of sets, their unions and intersections, *Zeitschrift für Wahrscheinlichkeitstheorie und verwandte Gebiete* **43**, 183–185.

A. V. Aho, J. E. Hopcroft, and J. D. Ullman (1974), *The Design and Analysis of Computer Algorithms*, Addison-Wesley, Reading, MA.

J. Akiyama, G. Exoo, and F. Harary (1981), Covering and packing in graphs IV: Linear arboricity, *Networks* **11**, 69–72.

N. Alon (1986a), Eigenvalues, geometric expanders, sorting in rounds and Ramsey Theory, *Combinatorica* **6**, 207–219.

——— (1986b), Eigenvalues and Expanders, *Combinatorica* **6**, 83–96.

——— (1988), The linear arboricity of graphs, *Israel Journal of Mathematics* **62**, 311–325.

——— (1990a), Transversal numbers of uniform hypergraphs, *Graphs and Combinatorics* **6**, 1–4.

——— (1990b), The maximum number of Hamiltonian paths in tournaments, *Combinatorica*, **10**, 319–324.

N. Alon, L. Babai, and A. Itai (1986), A fast and simple randomized parallel algorithm for the maximal independent set problem, *Journal of Algorithms* **7**, 567–583.

N. Alon and Z. Bregman (1988), Every 8-uniform 8-regular hypergraph is 2-colorable, *Graphs and Combinatorics* **4**, 303–305.

N. Alon and R. B. Boppana (1987), The monotone circuit complexity of Boolean functions, *Combinatorica* **7**, 1–22.

N. Alon and F. R. K. Chung (1988), Explicit construction of linear sized tolerant networks, *Discrete Mathematics* **72**, 15–19.

N. Alon, P. Frankl, and V. Rödl (1985), Geometrical realization of set systems and probabilistic communication complexity, *Proceedings of the 26th FOCS*, IEEE, New York, 277–280.

N. Alon, O. Goldreich, J. Hastad, and R. Peralta (1990), Simple constructions of almost k-wise independent random variables, *Proceedings of the 31st FOCS*, St. Louis, MO, IEEE, New York, 544–553.

N. Alon and N. Linial (1989), Cycles of length 0 modulo k in directed graphs, *Journal of Combinatorial Theory, Series B* **47**, 114–119.

N. Alon and P. Frankl (1985), The maximum number of disjoint pairs in a family of subsets, *Graphs and Combinatorics* **1**, 13–21.

N. Alon and D. J. Kleitman (1990), Sum-free subsets, In A. Baker, B. Bollobás, and A. Hajnál, eds., *A Tribute to Paul Erdős*, Cambridge University Press, 13–26.

N. Alon and V. D. Milman (1984), Eigenvalues, expanders and superconcentrators, *Proceedings of the 25th Annual FOCS*, Singer Island, FL, IEEE, New York, 320–322.
See also N. Alon and V. D. Milman (1985), λ_1, isoperimetric inequalities for graphs and superconcentrators, *Journal of Combinatorial Theory, Ser. B* **38**, 73–88.

A. E. Andreev (1985), On a method for obtaining lower bounds for the complexity of individual monotone functions, *Doklady Akademii Nauk SSSR* **282**:5, 1033–1037 (in Russian). English translation in *Soviet Mathematics Doklady* **31**:3, 530–534.

——— (1988), On a method for obtaining more than quadratic lower bounds for the complexity of π-schemes, preprint.

I. Bárány and Z. Füredi (1987), Empty simplices in Euclidean Spaces, *Canadian Mathematics Bulletin* **30**, 436–445.

J. Beck (1978), On 3-chromatic hypergraphs, *Discrete Mathematics* **24**, 127–137.

——— (1981), Roth's estimate of the discrepancy of integer sequences is nearly optimal, *Combinatorica* **1**, 319–325.

——— (1991), An algorithmic approach to the Lovász Local Lemma I, *Random Structures and Algorithms* **2**, 343–365.

J. Beck and T. Fiala (1981), Integer-making theorems, *Discrete Applied Mathematics* **3**, 1–8.

S. N. Bernstein (1912), Démonstration du théorème de Weierstrass fondée sur le calcul des probabilités, *Communications of the Society of Mathematics Kharkov* **13**, 1–2.

N. Blum (1984), A Boolean function requiring $3n$ network size, *Theoretical Computer Science* **28**, 337–345.

B. Bollobás (1965), On generalized graphs, *Acta Mathematica of the Academy of Sciences, Hungary* **16**, 447–452.

——— (1985), *Random Graphs*, Academic, New York.

——— (1988), The chromatic number of random graphs, *Combinatorica* **8**, 49–55.

B. Bollobás and P. Erdős (1976), Cliques in Random Graphs, *Mathematics Proceedings of the Cambridge Philosophical Society* **80**, 419–427.

R. B. Boppana and J. H. Spencer (1989), A useful elementary correlation inequality, *Journal of Combinatorial Theory, Series A* **50**, 305–307.

L. M. Brégman (1973), Some properties of nonnegative matrices and their permanents, *Soviet Mathematics Doklady* **14**, 945–949.

H. Chernoff (1952), A measure of the asymptotic efficiency for tests of a hypothesis based on the sum of observations, *Annals of Mathematical Statistics* **23**, 493–509.

F. R. K. Chung and R. Graham (1990) Quasi-random hypergraphs, *Random Structures and Algorithms* **1**, 105–124.

F. R. K. Chung, R. L. Graham, and R. M. Wilson (1989), Quasi-random graphs, *Combinatorica* **9**, 345–362.

A. Cohen and A. Wigderson (1990), Multigraph amplification, manuscript.

L. Danzer and B. Grünbaum (1962), Über zwei Probleme bezüglich konvexer Körper von P. Erdős und von V. L. Klee, *Mathematische Zeitschrift* **79**, 95–99.

R. M. Dudley (1978), Central limit theorems for empirical measures, *Annals of Probability* **6**, 899–929.

P. Erdős (1947), Some remarks on the theory of graphs, *Bulletin of the American Mathematics Society* **53**, 292–294.

——— (1956), Problems and results in additive number theory, *Colloque sur le Théorie des Nombres (CBRM, Bruselles)*, 127–137.

——— (1959), Graph theory and probability, *Canadian Journal of Mathematics* **11**, 34–38.

——— (1962), On circuits and subgraphs of chromatic graphs, *Mathematika* **9**, 170–175

——— (1963a), On a problem of graph theory, *Math. Gaz.* **47**, 220–223.

——— (1963b), On a combinatorial problem, I, *Nordisk Mat. Tidskrift* **11**, 5–10.

——— (1964), On a combinatorial problem, II, *Acta Mathematica of the Academy of Sciences, Hungary* **15**, 445–447.

——— (1965a), Extremal problems in number theory, *Proceedings of the Symposium on Pure Mathematics*, **VIII**, AMS, 181–189.

——— (1965b), On extremal problems of graphs and generalized graphs, *Israel Journal of Mathematics* **2**, 189–190

P. Erdős and Z. Füredi (1983), The greatest angle among n points in the d-dimensional Euclidean space, *Annals of Discrete Mathematics* **17**, 275–283.

P. Erdős and H. Hanani (1963), On a limit theorem in combinatorial analysis, *Publ. Math. Debrecen* **10**, 10–13.

P. Erdős and M. Kac (1940), The Gaussian law of errors in the theory of additive number theoretic functions, *American Journal of Mathematics* **62**, 738–742.

P. Erdős and L. Lovász (1975), Problems and results on 3-chromatic hypergraphs and some related questions, in A. Hajnal et al., eds., *Infinite and Finite Sets*, North-Holland, Amsterdam, 609–628.

P. Erdős and J. W. Moon (1965), On sets of consistent arcs in a tournament, *Canadian Mathematics Bulletin* **8**, 269–271.

P. Erdős and A. Rényi (1960), On the evolution of random graphs, *Magyar Tud. Akad. Mat. Kut. Int. Közl* **5**, 17–61.

P. Erdős and J. L. Selfridge (1973), On a combinatorial game, *Journal of Combinatorial Theory, Series A* **14**, 298–301.

P. Erdős and J. Spencer (1990), Lopsided Lovász local lemma and Latin transversals, *Discr. Appl. Math.* **30**, 151–154.

P. Erdős and P. Tetali (1990), Representations of integers as the sum of k terms, *Random Structures and Algorithms* **1**, 245–261.

R. Fagin (1976), Probabilities in finite models, *Journal of Symbolic Logic* **41**, 50–58.

C. M. Fortuin, P. W. Kasteleyn, and J. Ginibre (1971), Correlation inequalities on some partially ordered sets, *Commun. Math. Physics* **22**, 89–103.

P. Frankl and R. M. Wilson (1981), Intersection theorems with geometric consequences, *Combinatorica* **1**, 357–368.

P. Frankl, V. Rödl, and R. M. Wilson (1988), The number of submatrices of given type in a Hadamard matrix and related results, *Journal of Combinatorial Theory, Series B* **44**, 317–328.

M. Furst, J. Saxe, and M. Sipser (1984), Parity, circuits and the polynomial hierarchy, *Mathematical Systems Theory* **17**, 13–27.

Y. V. Glebskii, D. I. Kogan, M. I. Liagonkii, and V. A. Talanov (1969), Range and degree of realizability of formulas the restricted predicate calculus, *Cybernetics* **5**, 142–154 (Russian original: *Kibernetica* **5**, 17–27).

R. L. Graham and J. H. Spencer (1971), A constructive solution to a tournament problem, *Canadian Mathematics Bulletin* **14**, 45–48.

R. L. Graham, B. L. Rothschild, and J. H. Spencer (1990), *Ramsey Theory*, 2nd ed., Wiley, New York.

H. Halberstam and K. F. Roth (1983), *Sequences*, 2nd ed., Springer-Verlag, Berlin.

M. Hall (1986), *Combinatorial Theory*, 2nd ed., Wiley, New York.

L. Harper (1966), Optimal numberings and isoperimetric problems on graphs, *Journal of Combinatorial Theory* **1**, 385–394.

T. E. Harris (1960), A lower bound for the critical probability in a certain percolation process, *Mathematics Proceedings of the Cambridge Philosophical Society* **56**, 13–20.

J. Hastad (1988), Almost optimal lower bounds for small depth circuits, in S. Micali, ed., *Advances in Computer Research*, Volume 5: *Randomness and Computation*, JAI Press, Greenwich, CT, 143–170.

D. Haussler and E. Welzl (1987), ϵ-nets and simplex range queries, *Discrete and Computational Geometry* **2**, 127–151.

S. Janson (1990), Poisson approximation for large deviations, *Random Structures and Algorithms* **1**, 221–230.

A. Joffe (1974), On a set of almost deterministic k-independent random variables, *Annals of Probability* **2**, 161–162.

J. Kahn (1990), Coloring nearly-disjoint hypergraphs with $n + o(n)$ colors, to appear.

R. M. Karp (1990), The transitive closure of a random digraph, *Random Structures and Algorithms* **1**, 73–94.

R. M. Karp and V. Ramachandran (1990), Parallel algorithms for shared memory machines, in J. Van Leeuwen, ed., *Handbook of Theoretical Computer Science*, Volume A, Elsevier, New York, Chapter 17, 871–941.

M. Katchalski and A. Meir (1988), On empty triangles determined by points in the plane, *Acta Mathematica Hungary* **51**, 323–328.

G. O. H. Katona (1972), A simple proof of the Erdős Ko Rado theorem, *J. Combinatorial Theory, Series B* **13**, 183–184.

V. M. Khrapchenko (1971), A method of determining lower bounds for the complexity of Π-schemes, *Matematischi Zametki* **10**:1, 83–92 (in Russian). English translation in *Mathematical Notes of the Academy of Sciences of the USSR* **11** (1972), 474–479.

D. J. Kleitman (1966a), On a combinatorial problem of Erdős, *Journal of Combinatorial Theory* **1**, 209–214.

——— (1966b), Families of non-disjoint subsets, *Journal of Combinatorial Theory* **1**, 153–155.

J. Komlós, J. Pintz, and E. Szemerédi (1982), A lower bound for Heilbronn's problem, *Journal of the London Mathematics Society 2* **25**, 13–24.

M. Krachmer and A. Wigderson (1990), Monotone circuits for connectivity require super-logarithmic depth, *SIAM Journal of Discrete Mathematics* **3**, 255–265.

L. Lovász, J. Spencer, and K. Vesztergombi (1986), Discrepancy of set systems and matrices, *European Journal of Combinatorics* **7**, 151–160.

A. Lubotzky, R. Phillips, and P. Sarnak (1986), Explicit expanders and the Ramanujan conjectures, *Proceedings of the 18th ACM STOC*, ACM Press, New York, 240–246.
See also A. Lubotzky, R. Phillips, and P. Sarnak (1988), Ramanujan graphs, *Combinatorica* **8**, 261–277.

T. Luczak (1990), Component behavior near the critical point of the random graph process, *Random Structures and Algorithms* **1**, 287–310.

F. J. MacWilliams and N. J. A. Sloane (1977), The Theory of Error Correcting Codes, North-Holland, Amsterdam.

P. Mani-Levitska and J. Pach (1988), Decomposition problems for multiple coverings with unit balls, manuscript.

G. A. Margulis (1973), Explicit constructions of concentrators, *Problemy Peredachi Informatsii* **9**, 71–80 (in Russian). English translation in *Problems of Information Transmission* **9**, 325–332.

——— (1988), Explicit group-theoretical constructions of combinatorial schemes and their application to the design of expanders and superconcentrators, *Problemy Peredachi Informatsii* **24**, 51–60 (in Russian). English translation in *Problems of Information Transmission* **24**, 39–46.

J. Marica and J. Schonheim (1969), Differences of sets and a problem of Graham, *Canadian Mathematics Bulletin* **12**, 635–637.

D. W. Matula (1976), *The Largest Clique Size in a Random Graph*, Technical Report, Southern Methodist University, Dallas.

D. W. Matula (1981), Determining edge connectivity in $O(nm)$, *Proc. 28th Annual FOCS*, Los Angeles, CA, IEEE, 249–251.

B. Maurey (1979), Construction de suites symétriques, *Comptes Rendu Academie des Sciences Paris* **288**, 679–681.

V. D. Milman and G. Schechtman (1986), *Asymptotic Theory of Finite Dimensional Normed Spaces*, Lecture Notes in Mathematics 1200, Springer Verlag, New York.

J. W. Moon (1968), *Topics on Tournaments*, Holt, Reinhart & Winston, New York.

A. Nakayama and B. Peroche (1987), Linear arboricity of digraphs, *Networks* **17**, 39–53.

J. Naor and M. Naor (1990), Small-bias probability spaces: efficient constructions and applications, *Proceedings of the 22nd annual ACM STOC*, ACM Press, New York, 213–223.

A. Nilli (1990), On the second eigenvalue of a graph, *Discrete Mathematics*, in press.

J. Pach and G. Woeginger (1990), Some new bounds for epsilon-nets, *Proceedings of the 6th Annual Symposium on Computational Geometry*, Berkeley, CA, ACM Press, New York, 10–15.

R. Paturi and J. Simon (1984), Probabilistic communication complexity, *Proceedings of the 25th FOCS*, IEEE, New York, 118–126.

W. J. Paul (1977), A $2.5n$ lower bound on the combinational complexity of Boolean functions, *SIAM Journal on Computing* **6**:3, 427–443.

M. Pinsker (1973), On the complexity of a concentrator, *7th International Teletraffic Conference*, Stockholm, 318/1–318/4.

N. Pippenger and J. Spencer (1989), Asymptotic behaviour of the chromatic index for hypergraphs, *J. Combinatorial Theory, Series A* **51**, 24–42.

M. O. Rabin (1980), Probabilistic algorithms for testing primality, *Journal of Number Theory* **12**, 128–138.

P. Raghavan (1988), Probabilistic construction of deterministic algorithms: approximating packing integer programs, *Journal of Computer and Systems Sciences* **37**, 130–143.

F. P. Ramsey (1930), On a problem of formal logic, *Proceedings of the London Mathematics Society* **30**(2), 264–286.

A. A. Razborov (1985), Lower bounds on the monotone complexity of some Boolean functions, *Doklady Akademii Nauk SSSR* **281**:4, 798–801 (in Russian). English translation in *Soviet Mathematics Doklady* **31**, 354–357.

——— (1987), Lower bounds on the size of bounded depth networks over a complete basis with logical addition, *Mat. Zametki* **41**:4, 598–607 (in Russian). English translation in *Mathematical Notes of the Academy of Sciences of the USSR* **41**:4, 333–338.

V. Rödl (1985), On a packing and covering problem, *European Journal of Combinatorics* **5**, 69–78.

N. Sauer (1972), On the density of families of sets, *Journal of Combinatorial Theory, Series A* **13**, 145–147.

A. Schrijver (1978), A short proof of Minc's conjecture, *Journal of Combinatorial Theory, Series A* **25**, 80–83.

E. Shamir and J. Spencer (1987), Sharp concentration of the chromatic number in random graphs $G_{n,p}$, *Combinatorica* **7**, 121–130.

J. Shearer (1985), On a problem of Spencer, *Combinatorica* **5**, 241–245.

S. Shelah and J. Spencer (1988), Zero–one laws for sparse random graphs, *Journal of the American Mathematics Society* **1**, 97–115.

L. A. Shepp (1982), The XYZ-conjecture and the FKG-inequality, *Annals of Probability* **10**, 824–827.

R. Smolensky (1987), Algebraic methods in the theory of lower bounds for Boolean circuit complexity, *Proceedings of the 19th Annual ACM STOC*, ACM Press, New York, 77–82.

J. Spencer (1977), Asymptotic lower bounds for Ramsey functions, *Discrete Mathematics* **20**, 69–76.

——— (1985), Six standard deviations suffice, *Transactions of the American Mathematics Society* **289**, 679–706.

——— (1985), Probabilistic methods, *Graphs and Combinatorics* **1**, 357–382.

——— (1987), *Ten Lectures on the Probabilistic Method*, SIAM, Philadelphia.

——— (1990a), Threshold functions for extension statements, *Journal of Combinatorial Theory, Series A* **53**, 286–305.

——— (1990b), Counting Extension, *Journal of Combinatorial Theory, Series A* **55**, 247–255.

B. A. Subbotovskaya (1961), Realizations of linear functions by formulas using $+, \cdot, -$, *Doklady Akademii Nauk SSSR* **136:3**, 553–555 (in Russian).
English translation in *Soviet Mathematics Doklady* **2**, 110–112.

W. C. Suen (1990), A correlation inequality and a Poisson limit theorem for nonoverlapping balanced subgraphs of a random graph, *Random Structures and Algorithms* **1**, 231–242.

T. Szele (1943), Kombinatorikai vizsgálatok az irányitott teljes gráffal kapcsolatban, *Matematicko Fizicki Lapok* **50**, 223–256.
(For a German translation, see *Publ. Math. Debrecen* **13** (1966), 145–168.)

R. M. Tanner (1984), Explicit construction of concentrators from generalized N-gons, *SIAM Journal of Algebraic Discrete Methods* **5**, 287–293.

R. E. Tarjan (1983), *Data Structures and Network Algorithms*, SIAM, Philadelphia.

A. Thomason (1987), Pseudo-random graphs, in M. Karonski, ed., *Proceedings of Random Graphs, Poznán 1985*, *Annals of Discrete Mathematics* **33**, 307–331.

P. Turán (1934), On a theorem of Hardy and Ramanujan, *Journal of the London Mathematics Society* **9**, 274–276.

——— (1941), On an extremal problem in graph theory, *Matematicko Fizicki Lapok* **48**, 436–452.

V. N. Vapnik and A. Ya Chervonenkis (1971), On the uniform convergence of relative frequencies of events to their probabilities, *Theory of Probability Applications* **16**, 264–280.

W. F. de la Vega (1983), On the maximal cardinality of a consistent set of arcs in a random tournament, *Journal of Combinatorial Theory, Series B* **35**, 328–332.

I. Wegener (1987), *The Complexity of Boolean Functions*, Wiley–Teubner, New York.

A. Weil (1948), Sur les courbes algébriques et les variéstés qui sèn déduisent, *Actualités Scientifiques et Industrielles*, No. 1041.

J. G. Wendel (1962), A problem in geometric probability, *Mathematics Scandinavia* **11**, 109–111.

E. M. Wright (1977), The number of connected sparsely edged graphs, *Journal of Graph Theory* **1**, 317–330.

A. C. Yao (1985), Separating the polynomial-time hierarchy by oracles, *Proceedings of the 26th Annual IEEE Symposium on Foundations of Computer Science*, IEEE, New York, 1–10.

Index